Metacolloids in
Endogenic Deposits

Monographs in Geoscience

General Editor: Rhodes W. Fairbridge

Department of Geology, Columbia University, New York City

B. B. Zvyagin
Electron-Diffraction Analysis of Clay Mineral Structures–1967

E. I. Parkhomenko
Electrical Properties of Rocks–1967

A. I. Perel'man
The Geochemistry of Epigenesis–1967

L. M. Lebedev
Metacolloids in Endogenic Deposits–1967

In preparation:

A. S. Povarennykh
Crystal-Chemical Classification of Mineral Species

S. J. Lefond
Handbook of World Salt Resources

Metacolloids in Endogenic Deposits

Lev M. Lebedev

Institute of Geology of Ore Deposits, Petrography,
Mineralogy, and Geochemistry
Academy of Sciences of the USSR, Moscow

Translated from Russian by

John B. Southard

Department of Geology and Geophysics
Massachusetts Institute of Technology
Cambridge, Massachusetts

 Springer Science+Business Media, LLC 1967

Born in 1925 into the family of a mining engineer, L. M. Lebedev was graduated from Moscow State University in 1950. He has participated in many expeditions to all parts of the Soviet Union, from Transcaucasia to the Kurile Islands, and data collected from the mineral deposits of the many different regions visited form the basis for this volume.

Presently L. M. Lebedev is associated with the Institute of Geology of Ore Deposits, Petrography, Mineralogy, and Geochemistry of the Academy of Sciences of the USSR, and is engaged in research connected with hydrothermal processes and their relation to mineral and ore formation.

The original Russian text was published for the Siberian Branch of the Volcanological Institute of the Academy of Sciences of the USSR by Nauka Press, Moscow, in 1965.

Лев Михайлович Лебедев

Метаколлоиды в эндогенных месторождениях

METAKOLLOIDY V ENDOGENNYKH MESTOROZHDENIYAKH

METACOLLOIDS IN ENDOGENIC DEPOSITS

Library of Congress Catalog Card Number 65-25241

ISBN 978-1-4899-5614-9   ISBN 978-1-4899-5612-5 (eBook)
DOI 10.1007/978-1-4899-5612-5

© 1967 Springer Science+Business Media New York
Originally published by Plenum Press in 1967.
Softcover reprint of the hardcover 1st edition 1967

Preface

At the present time, many investigators acknowledge the well-known role of colloids in the formation of endogenic deposits, but they view the nature and extent of participation of colloids in ore deposition in various ways. Some suggest and to some extent demonstrate a possible importance of colloidal solutions in transporting mineral components; others deny this but acknowledge the possible existence of considerable quantities of a colloidal phase, in the form of a gel, in ore deposition. In addition, a number of workers deny that colloids play any part in ore deposition.

The importance of colloids in the formation of various kinds of mineral aggregates is confirmed by the existence in nature of mineral gels of various compositions. Study of these gels leads to the discovery of direct textural and morphological evidence of colloidal origin of mineral aggregates, and, by comparing these with the morphological and textural-structural features of metacolloidal aggregates, it may be possible to derive objective criteria to clarify the nature of these metacolloidal aggregates.

The book consists of two parts. The first part considers certain questions of the developmental history of metacolloidal aggregates and other superficially similar mineral aggregates. The second part is a description of the mineralogy of deposits with typical metacolloidal ores and also of the Pauzhetka deposit, which contains considerable accumulations of silica gel. This second part also outlines experimental data on the diagenesis of zinc sulfide and lead sulfide gels.

A number of my ideas are more or less open to debate, and should not be considered unambiguous and commonly accepted.

In working on the manuscript I profited by the valuable advice of V. S. Myasnikov, A. B. Gen'kin, and I. N. Kigai. Individual parts of this manuscript were kindly reviewed by F. B. Chukhrov and T. N. Shadlun. To all these persons I express my deep thanks. I also consider it necessary to make note of the high quality of the electron microphotography, which was done in the electron microscopy laboratory of the Institute of Geology of Ore Deposits, Petrography, Mineralogy, and Geochemistry, and I express my profound thanks to G. S. Gritsaenko, the director of the laboratory, and to his co-workers, A. I. Gorshkov and N. D. Samotoin.

Contents

PART II

Chapter 7

Chapter 8

Problems of the Developmental History of Metacolloidal and Superficially Similar Mineral Aggregates

At present, problems of the developmental history of mineral aggregates are among the most pressing problems of genetic mineralogy. Despite important advances in mineralogy both in the Soviet Union and abroad, our understanding in this field is less than scanty.

As metacolloidal mineral aggregates are widespread in endogenic ore deposits of various types and ages (from the end of the Precambrian to the present), problems of the developmental history of these aggregates are of first-order importance. They are undoubtedly more complex than similar problems for mineral aggregates of crystallization origin, because many morphological types of metacolloidal aggregates are similar in spite of diverse modes of origin.

The principal morphological types of metacollidal and superficially similar mineral aggregates are described below, mainly on the basis of the writer's own observations of the mineralogy of metacolloids in ore deposits and modern processes of formation and alteration of certain gangue mineral aggregates.

Chapter 1

Morphological Features of Mineral Aggregates and Their Genetic Significance

In a book devoted to metacolloidal aggregates it is impossible not to discuss certain general matters of the morphology of such aggregates as reniform, flow–deposited, collomorphic, and spherical aggregates. It should be borne in mind that many of these morphological types of mineral aggregates have long been considered to be characteristically of colloidal origin. The aggregates themselves have been considered to be metacolloids, i.e., formed by crystallization of a gel (Vernadskii, 1910, 1925; Betekhtin et al., 1958; Levitskii, 1953; Shadlun, 1942; Chukhrov, 1950, 1955; Radkevich, 1952; Wherry, 1914; Rogers, 1917; Boydell, 1925; Lindgren, 1933; Garrels, 1944). In the last decade some papers have appeared in which the attempt was made to explain the formation of the aggregates mentioned above by crystallization (Grigor'ev, 1949, 1953; Cherepanov, 1951). The theoretical positions taken in these works were from the standpoint of revision of the role of colloids in ore deposition (Grigor'ev, 1953, 1961). Without entering into a discussion of these positions (this will be done in the appropriate chapter), we note only that these workers attach almost no importance to colloids in the formation of these aggregates.

Thus, the genetic interpretation of such aggregates as reniform, spherical, collomorphic, and flow–deposited is not always unambiguous. Hence arises the necessity to seek additional criteria by which to determine unambiguously the genesis of these aggregates. Using some such criteria, forms such as various sorts of strictly flow–deposited forms are established, but others require the application of textural and structural data.

It should be especially emphasized that the problem of the relationship between form and texture is most acute in the study of spherical, reniform, and so-called "collomorphic" aggregates. Though texture

3

is in many cases direct evidence of genesis, it should not be forgotten that it is precisely the textures of the aggregates which are the most protean and subject to variation. On the other hand, shape as such is more conservative. In view of this, in deciding between colloidal or crystallization origin of these aggregates by analyzing their morphology and texture, one extremely important factor, time, must be taken into account. Gels cannot exist for an indefinitely long time; generally they become crystallized and recrystallized. The gel textures are thereby transformed into crystallization textures (granular, fibrous, radial, etc.), and the aggregate becomes a metacolloid.

Shape is more stable and retains its original features longer than do textures. Examples of this are numerous in the mineralogical literature: the formation of radial texture in recrystallization of oolites, recrystallization of single crystals, etc. The conservative nature of shape is very clearly manifested in recrystallization of aggregates in calcite stalactites. The numerous calcite stalactites in sinkholes in the Carboniferous limestones of the Moscow region are a good example. While preserving their ideal dripstone shape, these stalactites show considerable textural diversity, including various transitions from original concentrically zoned and finely

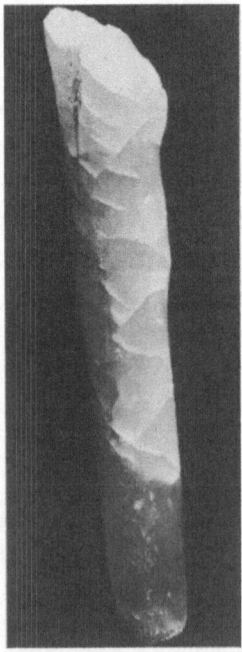

Fig. 1. Single-crystal calcite stalactite from the Carboniferous limestones of the Moscow region. The cleaved surface of the stalactite (shown by reflection) and the cleavage cracks along the rhombohedron are clearly visible (natural size).

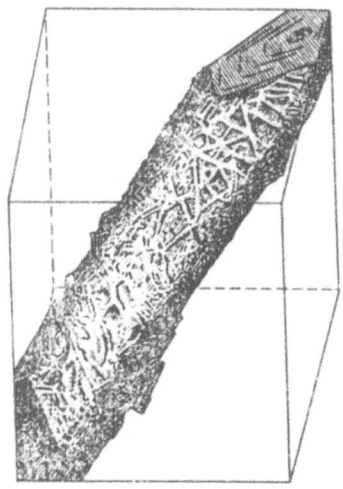

Fig. 2. Single-crystal galenite stalactite. Raibl deposit, Italy (after
Pošepný).

columnar aggregates, through radial and coarsely granular aggregates without zoning, to single crystals. When calcined, single-crystal
calcite stalactites show good rhombohedral cleavage (Fig. 1). The
single-crystal galenite stalactites described by Pošepný (1874) from
the Raibl lead—zinc deposit in Italy are very interesting in this
respect. As with the calcite stalactites mentioned above, these show
ideal cubic cleavage corresponding to a single crystal (Fig. 2).

Thus, the mere presence of crystallization texture is not evidence
of a crystallization origin of the aggregate, and, if the shape of the
aggregate shows features of colloidal origin, then it is preferable in
deciphering genesis to take morphological features into account.

The material below is a more or less detailed description of the
morphological and textural-structural features of mineral aggregates
that are most characteristic of metacolloids. Because the same
morphological varieties of minerals can be of different origin, the
characteristic genetic features of morphology and structure will be
analyzed in describing them.

Chapter 2

The Principal Structural-Textural Groups
of Spherical Mineral Aggregates

In studying the morphology of metacolloidal minerals (sphalerite, pyrite, cassiterite, chalcedony, opal, and others) the writer has observed a great diversity in structures and textures of spherical aggregates of these minerals. Analysis of the literature extends even further the range of mineral species that form spherical aggregates showing considerable diversity of structures and textures.

The existence of much factual material on spherical aggregates of various internal textures has made necessary a clear-cut classification of these aggregates on the basis of textures and structures and has raised the problem of clarifying the genetic significance of each mineral species. The meager data in the geological literature on classification of spherical aggregates are far from exhaustive and boil down mainly to a distinction between oolites and spherulites. Also, spherical aggregates that do not have well-defined internal texture are sometimes called pisolites. There is often confusion in terminology even in describing spherulites and oolites, despite sharp differences in structure.

However, neither spherulites nor oolites and pisolites exhaust the kinds of spherical aggregates that are observed in nature. More and more data on spherical aggregates of different types, which have been found in various metacolloidal deposits, have appeared in the recent mineralogical literature.

It is noteworthy first of all that many investigators have observed, in metacolloidal aggregates of ore minerals and gangue minerals, relict regions with globular texture, i.e., regions composed of very small spherical bodies (globules) which generally can be distinguished only at high magnifications. Grouped together, globules form larger spherical aggregates, which the writer distinguishes as globulites. Forms transitional between globulites and oolites are common.

Upon recrystallization, oolites pass over into spherulites by taking on a clearly expressed radiating structure, and so forth.

Study of the specific textural features of the aggregates mentioned above is of great importance in clarifying the conditions of formation both of the spherical aggregates themselves and of other mineral aggregates associated with them.

GLOBULES

Cases of globular texture of many metacolloidal minerals have been cited in a monograph by Chukhrov (1955). Globules are most characteristic of oxide and sulfide minerals. Globular texture in natural silica gel from solfataras of Golovnin volcano was noted by Naboko (Naboko and Sil'nichenko, 1957).

Electron-microscope study of silica gel deposited from one of the Pauzhetka hot springs allowed the particles of the gel and the character of the aggregates of these particles to be distinguished. The silica gel particles were globules, both individual and as chain aggregates (Fig. 3). A more detailed description is given in Chapter 10.

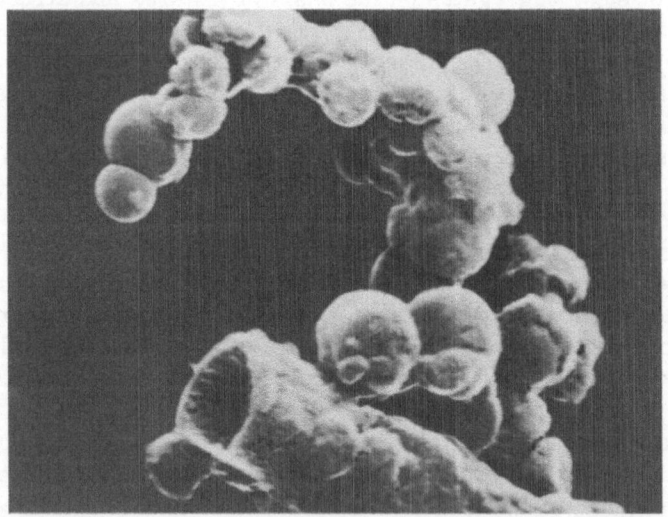

Fig. 3. Chain aggregate of silica gel globules on the surface of an SiO_2 membrane tube. Pauzhetka deposit, Second Teplyi Creek. Carbon replica (x 28,000).

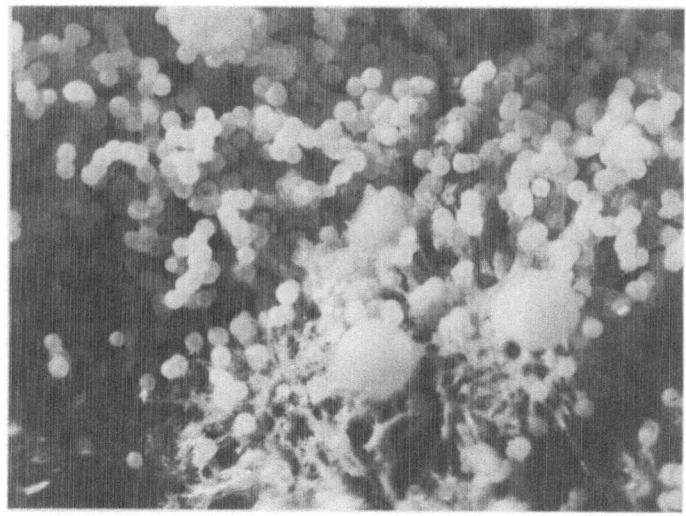

Fig. 4. Crystobalite macroglobules from cristobalite veins of the Pamach deposit, Georgia (x 30).

Andrushchenko (1954) has described, from the Polunochnoe deposit, opal consisting of accumulations of very fine globules (0.001 mm). Globular forms are also characteristic of opals of endogenic origin. The writer has observed exceptionally well-defined globular cristobalite aggregates in the Pamach chalcedony deposit (Fig. 4), where cristobalite forms thin (4 to 6 mm) veinlets in hydrothermally altered porphyritic tuff breccias. Unusual chain aggregates of opal globules have been noted in the opal—chalcedony veins of Karadag (the Crimea). Globular segregations of opal are common in chalcedony geodes from the Kertsy-Arach deposits of Armenia (Fig. 5).

Levitskii (1953) noted relics of globular texture of vein quartz associated with metacolloidal cassiterite from one of the Soviet Far East tin deposits. According to N. V. Petrovskii (personal communication), the quartz of one of the stages of mineralization of the Baleisk gold-ore deposit shows distinct globular texture.

Lebedev (1953) described globules of hydrohematite found as inclusions in quartz and calcite from the Beskempir iron-ore deposit (Fig. 6). A gel consistency of the matter in the central parts of large (0.5 mm) globules is characteristic of these globular forms.

The sealing-wax red coloring of the outer zones in quartz from Keremet-tas (Kazakhstan) is caused by inclusions of very fine hydrohematite globules.

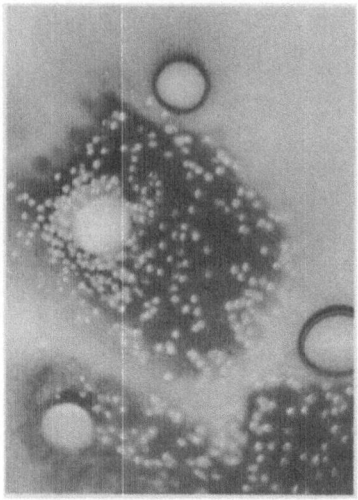

Fig. 5. Globular segregations of opal
in chalcedony. Kertsy-Arach deposit
(× 25).

Lebedev (1959a) observed relicts of globular texture of metacolloidal cassiterite among cryptocrystalline quartz—cassiterite aggregates of the Shakh-Shagaila deposit. Very small flakes of metacolloidal cassiterite are evenly distributed in the cryptocrystalline
quartz groundmass, grouped as poorly defined concentric zones. The
flaky particles, which adhere to one another, form denser spherical
bodies (globules). Besides accumulations in metacolloidal quartz,
cassiterite globules generally form the centers of cassiterite oolites.

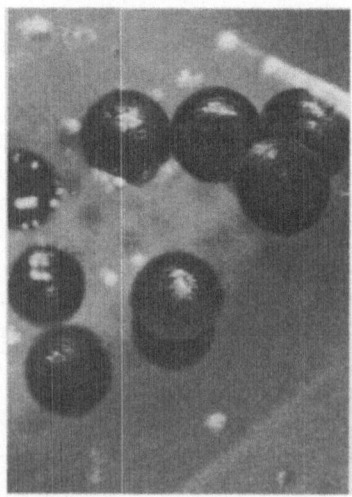

Fig. 6. Hydrohematite globules in
quartz. Beskempir deposit, Mangyshlak (× 150).

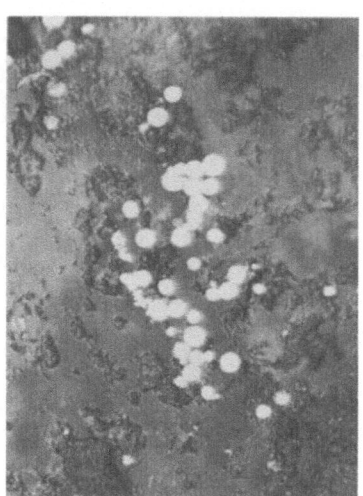

Fig. 7. Globular segregations of pyrite in very small pores of propylitized tuffs. Pauzhetka deposit, Kamchatka (x 185).

Globular texture of aggregates is characteristic also of meta-colloidal sulfides. Globular pyrite is widespread in both endogenic and sedimentary deposits (Fig. 7). It has been noted in the Iokun'zh, Aidyrla, Pauzhetka, and other deposits. Sterk (1953) noted globular texture in metacolloidal sphalerite (brunckite) from Cercapuquio, Peru. Study of the brunckite under an electron microscope in the electron-microscopy laboratory of the Institute of Geology of Ore Deposits, Petrography, Mineralogy, and Geochemistry (IGEM) of the

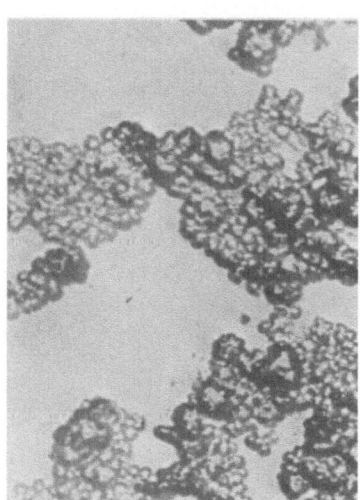

Fig. 8. Globular aggregate of sphalerite (brunckite), Cercapuquio, Peru. Carbon replica (x 24,000).

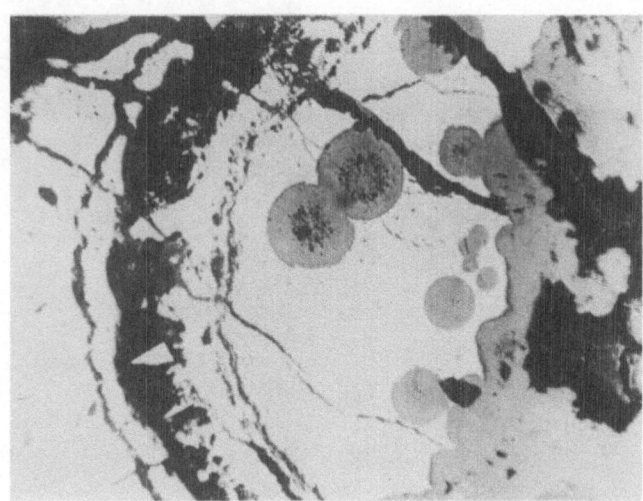

Fig. 9. Globular segregation of sphalerite in pyrite. Verkhnyaya-
Kvaisa deposit (x 150; after T. V. Ivanitskii).

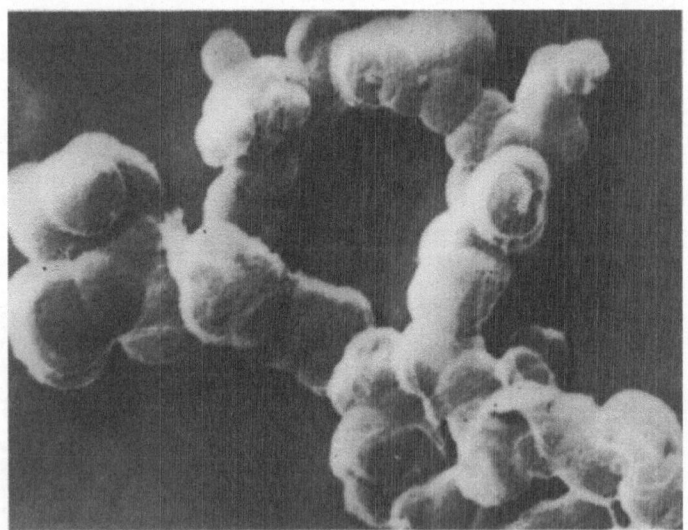

Fig. 10. Chain aggregate of allophane globules. Chernorechensk Cavern,
Moscow region. Carbon replica (x 35,000).

Academy of Sciences of the USSR confirms Sterk's data (Fig. 8). Ivanitskii (1953) observed relicts of globular texture in massive varieties of collomorphic sphalerite and pyrite from the Verkhnyaya-Kvaisa deposit (Fig. 9). Study of metacolloidal sphalerite from the Iokun'zh deposit (Lebedev, 1959a) has shown that morphologically this sphalerite is close to the Peruvian brunckite. Both continuous, massive varieties and pisolitic aggregates of it show globular texture.

In conclusion, it should be noted that, aside from the minerals mentioned above, globular textures are most often found in the allophanoid group. Figure 10 shows an electron photomicrograph of allophane forming an accumulation of white earthy masses in drusy cavities of quartz from Podol'sk (Moscow region). It is seen in the photograph that this allophane consists of very small, chain aggregates of globules.

Besides cases of globular texture of metacolloidal minerals that are observable in nature, many investigators have observed globular texture in gels of various compositions obtained in the laboratory.

Rozhkova obtained iron and aluminum oxides with globular texture* by slow, careful addition of an ammonia solution to solutions of salts of ferrous or ferric iron and aluminum. The iron and aluminum hydrates that precipitate in these conditions consist of accumulations of small, rounded forms that have the following unusual structure: The outer part is a rather thin, light-colored envelope, and the inner part consists of darker material. The pisolites† are light-colored immediately after they form and are at the surface (upon precipitation by ammonia), and then later, as they gradually adhere to one another and are covered by an overall envelope, they form larger pisolites,‡ which slowly settle to the bottom.

Rozhkova established that, independent of the reagent, these forms of iron and aluminum hydroxides come into being under a single indispensable condition, change in the pH of the medium. Rozhkova noted that, with a pH of 2.5 to 3, light, hollow, transparent "sheaths" are formed, which upon further addition of alkali become dense, opaque, and stable for a long time (at a pH of 4). Rozhkova also noted that the presence of colloidal silica in the "leguminous" forms of iron and aluminum hydroxides has a great effect upon their strength.

*Called "pisolitic texture" in Rozhkova's work (Rozhkova and Solov'ev, 1936). Rozhkova calls all spherical forms pisolitic.

†In this case, the gel clots and globules are formed.

‡That is, globulites, spherical aggregates composed of globules (see below).

The present writer has obtained globular zinc sulfide aggregates in the laboratory. It was established that freshly obtained zinc sulfide gels form globules during coagulation: The overall structureless mass of the gel separates into individual clots, at first friable, which later become more distinctly spherical. The gel takes on a granular texture. The individual textural units are very small spherical bodies—globules. The size of the globules in the zinc sulfide gels ranges from 0.1–0.3 μ to tenths of a millimeter. Figure 152 is an electron photomicrograph of a zinc sulfide gel. The globular texture of the gel, which was aged for six or seven months, shows clearly in the photograph. Upon longer aging, the texture of the gel becomes more complicated because of incipient crystallization and the formation of more complex spherical aggregates.

Berestneva (1953) described globular forms of arsenic sulfide ranging from several to 200 mμ in size. During subsequent diagenetic changes, the globules formed chain aggregates and larger spherical aggregates. Quinke (1902) traced the formation of colloidal silica globules. The present writer has observed both individual globules and chain aggregates of globules in electron-microscope study of two-component silica—stannous gels.

Watanabe (1924) obtained globular aggregates of lead and copper hydroxides. Electron-microscope study of sols of aluminum hydroxide and titanium dioxide by Berestneva, Koretskaya, and Kargin (1951) showed that the particles in sols of these compounds obtained in the laboratory at ordinary temperatures and normal atmospheric pressure were globules 100 mμ to 0.8 μ in size. Crystallization began after several hours. The globular titanium dioxide particles crystallized two hours after the sol was obtained. The aluminum hydroxide globules were more stable; they crystallized over two or three months.

Thus, globules are spherical forms with homogeneous gel consistency formed after coagulation of a gel.

The formation of globules begins immediately after the gel coagulates and is caused by the excess free surface energy at the boundaries between the two phases. As a rule, the globules are very small and show up in metacolloids only at the highest magnifications of optical microscopes. Electron microscopy shows more clearly the globular texture of gels and certain metacolloids.

During subsequent diagenesis of the gel, there is a tendency toward decrease in the degree of dispersion of the system. Coagulation leads either to a substantial increase in the size of the globules or to the formation of more complex aggregates, globulites, which are formed under conditions in which coalescence of globules has already become impossible.

Examples of globules made larger by coalescence are magnetite macroglobules from the Kezhemsk deposit, hydrohematite from the Beskempir deposit, cassiterite from the Shakh-Shagaila deposit, opal from the Pamach deposit, and sphalerite from the Verkhnyaya-Kvaisa deposit.

GLOBULITES

Besides continuous globular-granular texture of metacolloidal aggregates of minerals, so-called "thromboidal" aggregates (Chukhrov, 1955), which are spherical aggregates of globules, are often observed. Such aggregates, though spherical, are texturally more complex; the writer places them in a second textural group of spherical mineral aggregates.   It is proposed that aggregates of this type be called "globulites."

Globulites, because they are composed of globules, are characterized by well-defined globular texture.   The existing data allow us to distinguish two types of globular texture observed in globulites.

1. Uniformly globular: The globulite consists of an unordered aggregate of densely crowded globules. There is no regularity in the arrangement of the globules.

2. Chain-zonal globular: The globulite consists of globules that are grouped in concentric chains.

Globulites with uniformly globular texture have been found in pyrite from the Kerchensk deposit (Chukhrov, 1936) and the Aidyrla deposit (Fig. 11; Gritsaenko et al., 1950).   Chain-zonal pyrite globulites were also found in the latter deposit. The writer has observed pyrite globulites with uniformly globular texture in calcite–galenite veinlets of the Iokun'zh deposit and in voids in propylitized rocks in the region of the Pauzhetka hot springs (Fig. 12; Lebedev, 1959a, 1961). Malinovskii (1955 observed similar aggregates of pyrite as inclusions in phosphatic concretions in Podolia.

Globulites of similar texture are characteristic of sphalerite from the Iokun'zh deposit.   At high magnifications they are observed as group accumulations in a uniform globular sphalerite mass. The largest number of sphalerite globulites is generally in regions enriched in galenite (Fig. 13).   In experimental studies of the aging of ZnS gels and mixed ZnS + PbS gels, sphalerite globulites with uniformly globular texture were formed in the early stages of

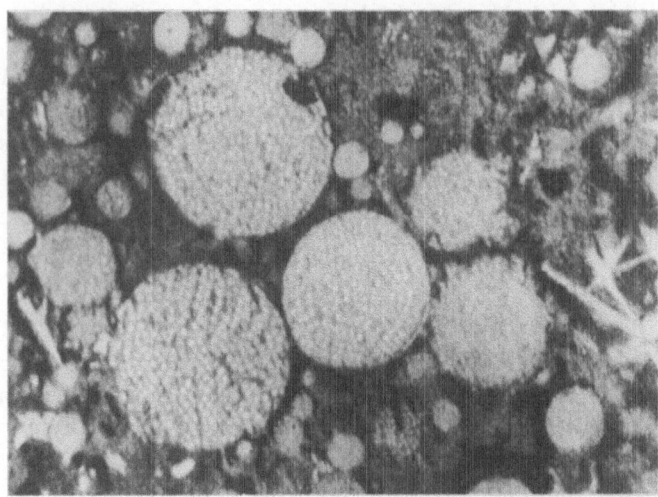

Fig. 11. Pyrite globulites with uniformly granular texture. Aidyrla
deposit (x 690; after G. S. Gritsaenko).

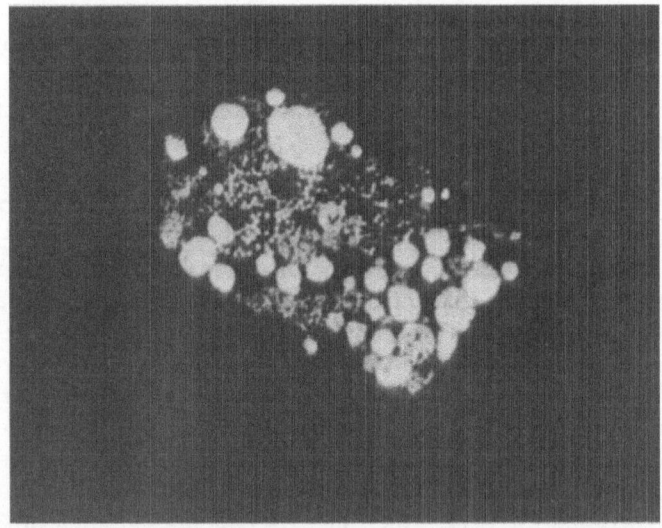

Fig. 12. Pyrite globulites and globules filling a leached cavity in
propylitized tuff. Pauzhetka deposit, Kamchatka (x 800).

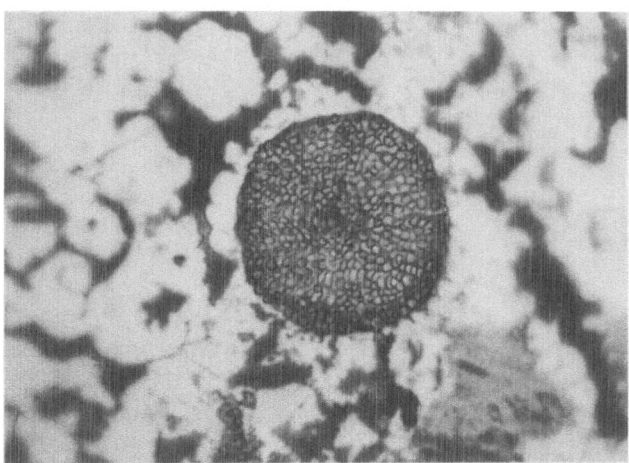

Fig. 13. Sphalerite globulite with uniformly granular texture. White,
galenite; black, pores. Iokun'zh deposit (× 800).

diagenesis of the gel. Also, as with natural metacolloidal aggregates
of sphalerite and galenite, in mixed ZnS + PbS gels considerable
accumulations of large sphalerite globulites were present in regions
enriched in the galenite component (see Fig. 167).

Uniformly globular texture is also noted for cristobalite globulites from the Pamach chalcedony deposit and globulites of natural
silica gel from Pauzhetka.

Chain-zonal texture of globulites is less common. This texture
is characteristic of pyrite globulites from Iokun'zh and is less characteristic of, but not uncommon in, spalerite globulites from the
Iokun'zh and Kvaisa deposits (Fig. 14). Volynskii (1946) noted pyrite
globulites with chain-zonal texture in the Aidyrla deposit (Fig. 15).
The writer has observed similar globulites of metacollidal cassiterite in cryptocrystalline quartz from the Shakh-Shagaila deposit.
Characteristic of the cassiterite globulites of this deposit are (1)
variable alternation of quartz and concentric globular chains of
metacolloidal cassiterite and (2) almost complete coalescence of
globules in the individual shells to form a uniform cassiterite zone
that loses its globular texture. Such globulites are transitional to
two-component (quartz—cassiterite) oolites.

The so-called pisolites often found in sedimentary iron ores
should probably also be classified as globulites. In all probability,
most of these pisolites are large globulites that have partly or completely lost their globular texture during subsequent diagenesis.

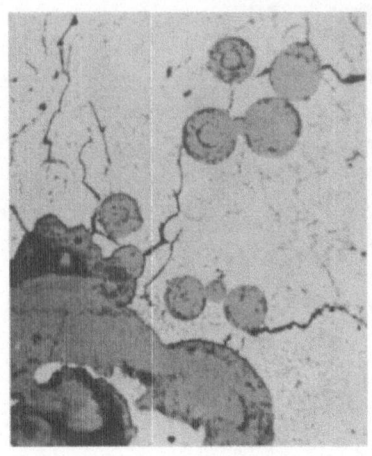

Fig. 14. Sphalerite globulites with signs of chain-zonal texture. Below, irregularly collomorphic sphalerite; white groundmass, pyrite. Truskovets deposit (× 150; after A. D. Genkin).

OOLITES

Oolites are spherical mineral aggregates characterized by distinct concentrically zoned structure. Oolites larger than 2 mm are conventionally called pisolites. That oolites are widespread among sedimentary mineral formations is well known. They are characteristic of mineral aggregates of hypogene origin to a lesser extent, but even in these aggregates they are not uncommon. Among hypogene aggregates, oolites are usually found both in ore metacolloids and in gangue metacolloids.

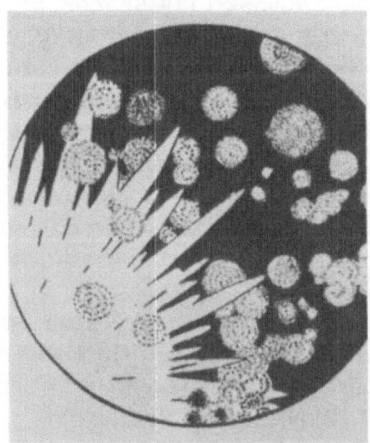

Fig. 15. Pyrite globulites with distinct chain-zonal texture. White acicular mineral, millerite; Aidyrla deposit (× 400; after I. S. Volynskii).

Morphology of Oolites

Though oolites are usually spheroidal (Fig. 16), various deviations from more or less strictly spherical shape are common. In most such cases the external shape of the aggregate is only slightly distorted and retains its spherical nature. Extreme distortions, even to the appearance of forms with faces (polygonal oolites), are also known.

Deviations from the ideal shape very often arise as early as during formation of the oolites. The envelopment of several linearly attached oolites by an overall envelope at some stage during growth leads ultimately to the development of ellipsoidal shapes (Fig. 17). The envelopment of differently oriented linear arrangements of small oolites leads to the development of extremely anomalous amoeba-shaped forms.

Unusual inherited shapes of oolites often develop. In this process, the shape of the growing oolites is determined by the size and shape of the rock or mineral fragment that forms the center of the oolite. If these fragments are small and the quantity of surrounding material is considerable, weakly deformed and generally spherical oolites are formed (Fig. 18). When the relations are reversed, i.e., when the fragments are large and the quantity of surrounding material is small, sharply anomalous shapes are formed—hemispherical compacted oolites and polygonal oolites of various shapes (Fig. 19).

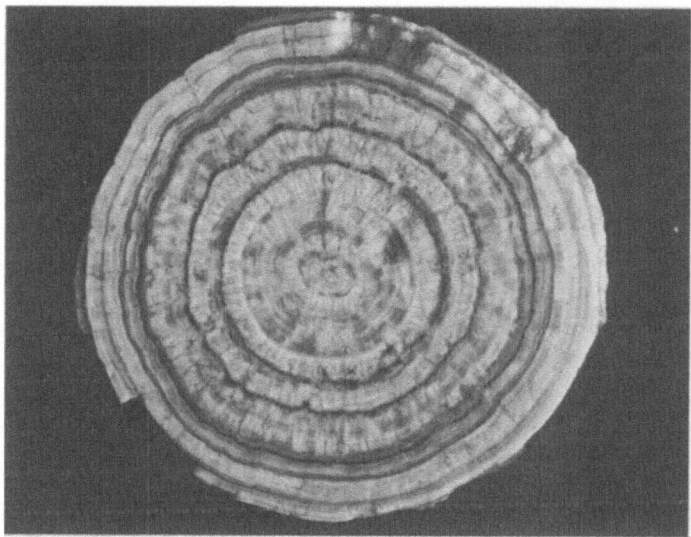

Fig. 16. Normally developed aragonite pisolite. Akhaltsikhe well (x 4).

Fig. 17. Abnormally developed aragonite pisolite. The distorted form is
related to the mantling of linear aggregate of oolites during growth.
Akhaltsikhe well (x 4).

The various types of deviation from the usual spherical shape
that develop during the formation of oolites have been observed in
many places by the writer in the zinc ores of the Iokun'zh deposit and
in aragonite oolites that are now being formed at hot springs of the
Akhaltsikhe well.   Calcite oolites and pisolites with inherited polyhedral shapes have been observed by Maksimovich (1955) in caverns of

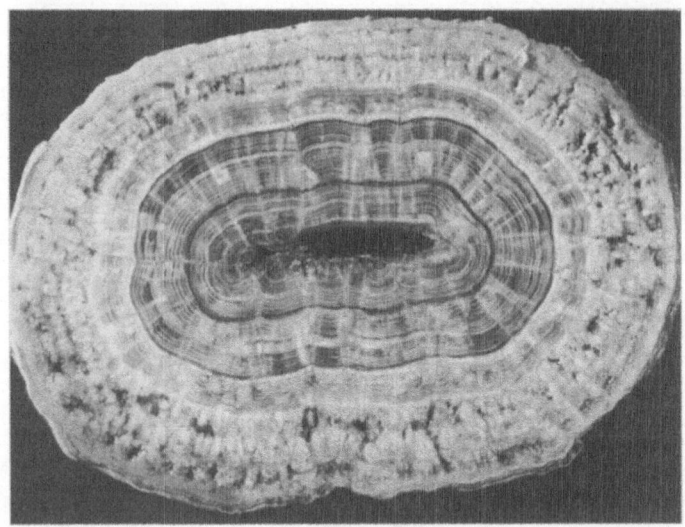

Fig. 18. Slightly distorted aragonite pisolite whose center is a small, flat,
elongated rock fragment (x 4).

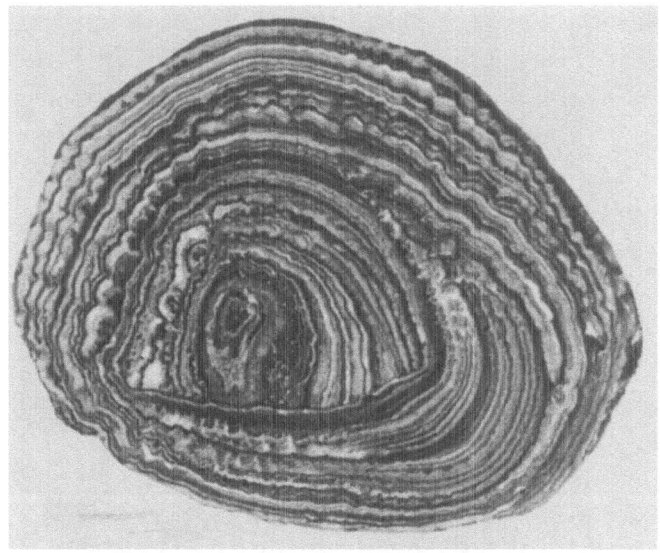

Fig. 19. Hemispherical sphalerite pisolite formed by the mantling of a
fragment of an earlier pisolite (x 3).

the Kizelov region and by Baker and Frostick (1951) in Angel Cave in
Australia.

Oolites also change shape after they form, during diagenesis.
It is noteworthy that oolites that form by crystallization do not undergo
any substantial changes during diagenesis. When the oolites formed in
colloidal media and when the matter that forms the oolites passed
through a gel stage, the oolites are distorted during diagenesis also.
In this case, even slight compaction of oolitic accumulations leads to
the appearance of flat surfaces at points of contact between oolites
and to the development of elements with polyhedral aspect among the
oolites (Fig. 20). Upon stronger compaction, the oolites take on a
distinct polyhedral appearance and acquire faces. Pavlov (1956)
described in detail magnetite oolites (pisolites) of this sort from the
Kezhemsk deposit. Besides the development of polyhedral forms,
upon appreciable compaction the oolites undergo stronger changes;
hollows of various kinds are formed, and often smaller oolites are
pressed into adjacent larger oolites. In this there is, so to speak, a
squeezing out of individual parts of the oolite, which takes on an
unusual asymmetrical palmate shape with a characteristic asymmetrical structure of the concentric shells. In some cases the oolites
are so strongly deformed that they lose the basic features of spherical
aggregates and are transformed into irregular collomorphic forms.

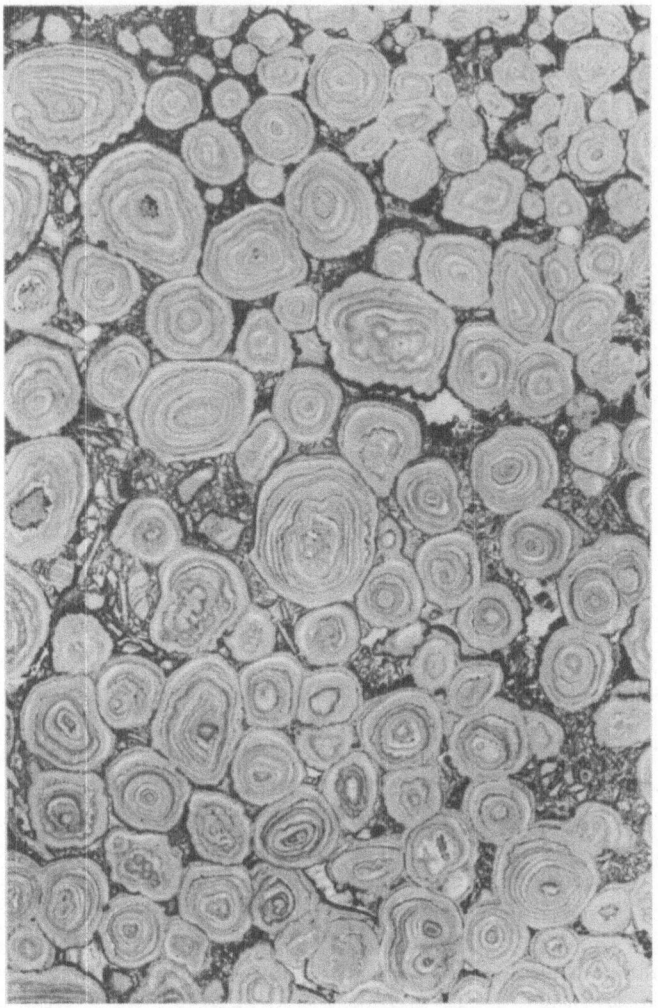

Fig. 20. Flat and deformed surfaces of sphalerite pisolites formed
during compaction of the aggregate. Iokun'zh deposit (natural size).

During compaction of oolitic accumulations the oolites are often
displaced relative to one another. When the material of the oolites
has retained a viscous consistency, the surfaces of the oolites take
on a distinctive corrugation that is related to the formation of micro-
plications in the outer shells (Fig. 21).

Finally, because ore gels have a high density, oolites which are
composed of these gels and which retain a viscous consistency for
some time must inevitably take on, under the influence of gravity, a

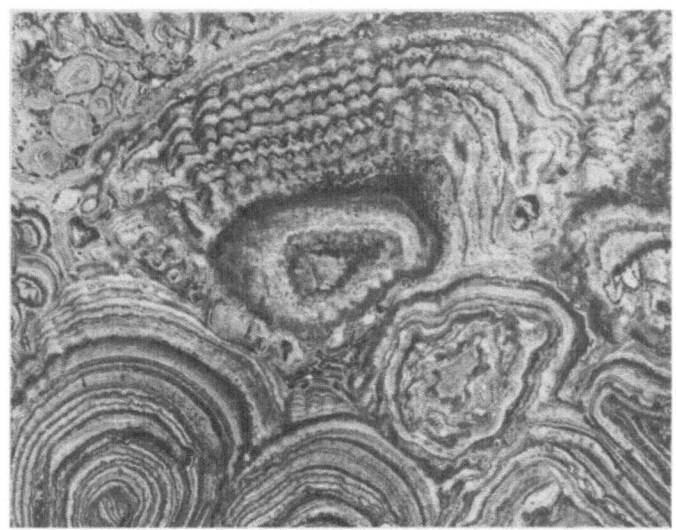

Fig. 21. Microplicated outer shells in sphalerite pisolites. Iokun'zh deposit (x 2).

shape whose asymmetry reflects the direction of gravity. The higher the density of the gel, the greater will be the deviations from spheroidal shape among the oolites. The writer has observed such changes of shape in sphalerite—galenite pisolites in the Iokun'zh deposit. Compacted and hemispherical forms predominate among these two-component pisolites (Fig. 22). Only in isolated instances are more or less regular spheroidal forms found among them.

Thus, all the morphological features of oolites as a whole reflect very clearly the features of the formation of oolites, and in many

Fig. 22. Hemispherical sphalerite—galenite pisolites with asymmetry caused by gravitational settling. Iokun'zh deposit (x 2).

cases these features permit an unambiguous decision about the
nature of the oolites.

Various diagenetic changes in shape are characteristic of oolites
of colloidal origin.   At the root of these changes is the viscosity
(plasticity) of the material forming the oolites and, in connection with
this, the widespread appearance of plastic deformations. Gravitational
flattening of oolites is also characteristic. In those formed by crystallization there is no change of shape by plastic deformation during
diagenesis.

Zoning of Oolites

Concentrically zoned structure in oolites is the result of rhythmic
processes during their formation.   Rhythmic processes, which are
widespread in nature and which give rise to stratification in sedimentary rocks and zoning of minerals and mineral aggregates, are
subdivided into external and internal rhythmic processes. Accordingly,
zoning in oolites is subdivided into two types depending upon the factor
of rhythmicity.

1. Depositional zoning is caused by external rhythmic processes.
Among such processes operating during the formation of oolites are
pulsating migration of solutions, seasonal changes in concentration
of solutions, release of carbon dioxide from the solutions upon agita-
tion of the solutions by drops (in cavern waters), and a number of
other processes.

Depositional zoning in oolites is characterized by sharp and distinct boundaries between shells, caused by the presence of parting
surfaces between the shells.   The clearest depositional zoning is
manifested in concentric parting of the oolites. This is most characteristic of oolites of crystallization origin. It appears less clearly
in oolites of colloidal origin. It is noteworthy that, in some carbonate
oolites in which the material of the concentric shells was deposited
initially as a gel but later crystallized rapidly, concentric parting
appears as clearly as in oolites formed by crystallization.

A characteristic example of depositional zoning is the zoning of
numerous carbonate oolites and pisolites that form in hot springs
(Akhaltsikhe, USSR, and Karlovy Vary, Czechoslovakia), in caverns
and abandoned mines (Carlsbad, New Mexico, and mines in Idaho;
also Australia and other regions) (Hess, 1929; Mackin, 1945), and in
lakes and marine basins (Vital', 1948; Brodskaya, 1954). Depositional
zoning is also characteristic of sphalerite pisolites from the Iokun'zh
deposit.

The attempt is often made to tell the age of oolites by the number of shells. Thus, Mathews (1930) assumed that each shell in the oolites of the Great Salt Lake corresponds to an annual period of growth. By counting the number of shells, Mackin (1945) found an age of 35 to 42 years for calcite pisolites forming in an abandoned mine in Idaho. However, it should be borne in mind that such computations are correct only when it has been demonstrated that the rhythmicity of formation of the shells is caused by strictly seasonal changes.

2. Diffusional zoning is caused by internal rhythmic processes. Among such processes in the formation of oolites are periodic precipitation of pigmenting material, caused by diffusion of electrolytes

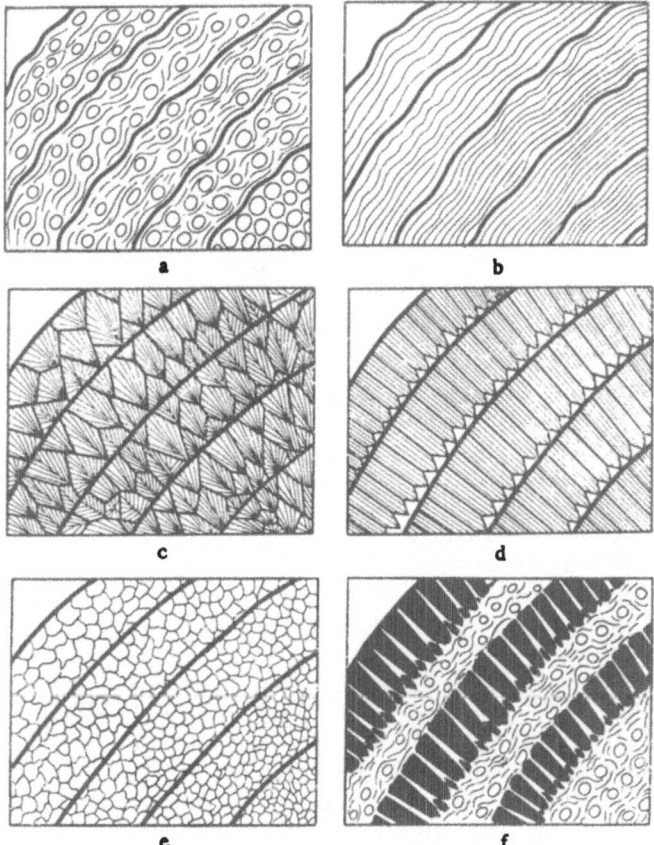

Fig. 23. Principal types of textures of oolite shells (sketches). (a) Globular-collomorphic; (b) collomorphic; (c) spherulitic; (d) columnar; (e) even-grained; (f) alternation of shells with different texture.

in the gels and by diffusional segregation of layers in a two-component gel. These processes are observed in oolites that form by coagulation or agglomeration of colloidal particles.

A certain vagueness and diffuseness of the boundaries between differently colored shells is characteristic of diffusional zoning. There are no parting surfaces between shells, and so concentric parting is absent.

Pavlov (1956) noted diffusional zoning in hypogene magnetite oolites and pisolites from iron ores of Eastern Siberia. Diffusional zoning is also characteristic of metacolloidal cassiterite oolites from the Shakh-Shagaila deposit. The galenite–sphalerite pisolites of the Iokun'zh deposit are a typical example of diffusional segregation of two-component oolites in the gel condition.

Zoning in oolites, because it reflects rhythmicity in formation, can serve as a certain criterion of the nature of the oolites in a number of cases. Thus, in the great majority of cases distinct concentric parting is evidence that the oolites originated by crystallization. On the other hand, diffusional zoning is characteristic of oolites of colloidal origin.

Texture of Oolites

In deciphering the nature of oolites it is of great genetic importance to determine the textures of the mineral aggregates that form each shell of the oolite. Analysis of information in the literature and study of the textures of the mineral aggregates forming oolites that the writer has observed permit the following most widespread textures to be distinguished (Fig. 23):

1. Globular-collomorphic
2. Collomorphic
3. Spherulitic
4. Columnar
5. Granular

Globular-collomorphic texture reflects the morphological features of particles that form both during coagulation of the sol and in the early stages of diagenesis of the gel. As a rule, when such aggregates are studied by the usual methods of optical microscopy, only large globules and globulites within festooned-collomorphic regions are observed. The latter are usually dense aggregates consisting of either extremely small globules or irregularly collomorphic agglomerations that are formed during coalescence of globules (Lebedev,

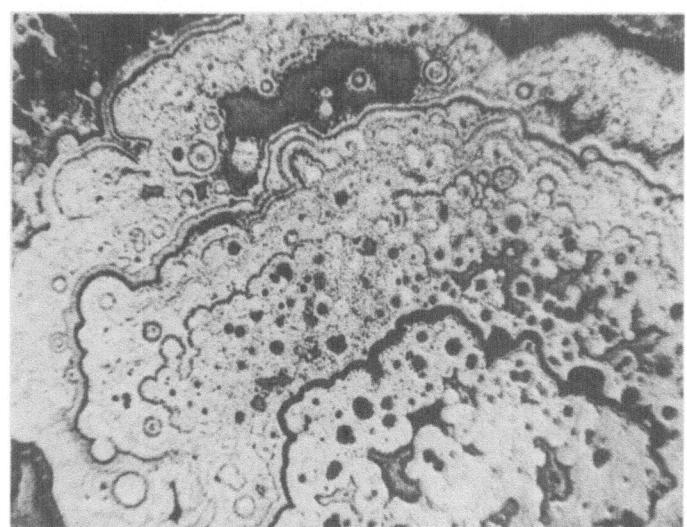

Fig. 24. Globular-collomorphic texture of shells of sphalerite oolite from
Truskovets (x 90; after A. D. Genkin).

1963). A characteristic example of globular-collomorphic texture is
that in sphalerite oolites from Truskovets (Fig. 24).

Collomorphic texture also reflects the morphological features of
the particles of the gel but corresponds to later stages of diagenesis
of the gel. Here, globular forms are noted in isolated cases as relicts.
The writer (Lebedev, 1959a, 1959b) has observed collomorphic shell
texture in sphalerite and galenite—sphalerite pisolites from Iokun'zh
and in cassiterite oolites from Shakh-Shagaila. Moreover, like globular-collomorphic texture, it is characteristic of sphalerite oolites of
the Truskovets deposit.

Spherulitic texture can arise during crystallization of a gel, in
the latest* stages of diagenesis. On the other hand, spherulitic shell
texture is apparently most characteristic of oolites formed by crystallization. In this case its formation can be related to the appearance
of spherulite nuclei on a spheroidal surface at the beginning of each
growth cycle of the oolite. The formation of a shell can thus be
considered to be group growth of spherulites according to the pattern
proposed by Grigor'ev (1961).

Spherulitic shell texture, even though it is a crystallization
texture, cannot be considered a sure criterion that the oolites have
formed by crystallization. Apparently, each time spherulitic shell

*For gels of various compositions this stage is not equivalent in a time sense (see p. 292).

Fig. 25. Spherulitic texture of shells of aragonite pisolite. Akhaltsikhe
well (x 46; crossed nicols).

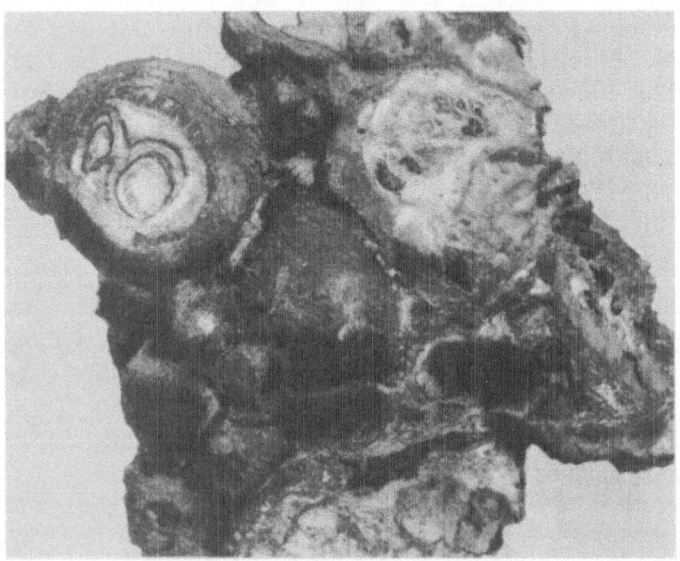

Fig. 26. Columnar texture of outer shells of galenite in galenite-
sphalerite pisolites. Iokun'zh deposit (natural size).

texture is observed, additional criteria must be used in deciding the genesis of the oolites.

Study of modern aragonite pisolites whose material passed through a gel stage (during deposition) has made it possible to distinguish several specific features of spherulitic texture arising by crystallization of a calcium carbonate gel. Among these features we should note the following:

1. The absence of clearly defined geometrical selection of spherulites in the shells, which in most cases consist of very fine, relatively freely developed spherulites. Spherical forms without radiating structure are often noted among these spherulites. In isolated cases the spaces between spherulites are filled with an isotropic carbonate mass or one of extremely weak polarization.

2. The presence of irregular cavities between the spherulites forming the shells (Fig. 25).

3. The deformation of the primary concentric, variously colored zones in the shells (or groups of shells), where nonetheless there is geometrical selection of spherulites. In this it is noteworthy that often there are regions composed of cryptocrystalline calcium carbonate without spherulitic texture in zones with geometrical selection of spherulites.

All this taken together is evidence that the spherulitic shell textures of the aragonite pisolites are secondary.

Columnar texture is most characteristic of oolites formed by crystallization. It should be noted, however, that this shell texture, too, is not an unambiguous criterion that the aggregates originated by crystallization. Often this texture is formed by crystallization of globular-collomorphic aggregates and by recrystallization of spherulitic aggregates. Columnar textures vary widely, depending upon the geometry of the component crystals, from acicular to coarsely columnar (Fig. 26). Columnar shell texture is characteristic of many carbonate, fibrous chlorite, and other oolites that form various sedimentary rocks.

It should be pointed out that the columnar textures discussed above, which occur without change within all the shells of the oolite, are often quite incorrectly interpreted to be radiating structure of the oolite. This interpretation contradicts the very nature of the oolite, because the formation of the shells, caused in this case by external rhythmic processes, is related to certain breaks in the deposition of material and thus to the nonsimultaneous formation of the mineral aggregates that form the various shells. "Pseudospherulitic" textures are not characteristic of oolites with diffusional

Fig. 27. Metacolloidal-granular texture of shells of sphalerite pisolites
from the Iokun'zh deposit. Carbon replica (× 11,000).

zoning that are composed entirely of cryptocrystalline metacolloidal
aggregates. Moreover, radiating structure is ordinarily formed in
oolites during the latest recrystallization, and concentrically zoned
structure then disappears. Rozhkova (Rozhkova and Solov'ev, 1937)
observed directly such recrystallization of carbonate oolites into
spherulites during experiments on the production of oolitic aggregates
of carbonates of Ca, Mg, Mn, Fe, and other elements.

Granular texture is less common in oolites than the above
textures. Depending upon the size of the crystals that form the
aggregates, two types of this texture should be recognized, metacolloidal and allotriomorphic.

Metacolloidal-granular texture corresponds to an even-grained
aggregate with crystal sizes ranging from 1 to $10\,\mu$. It is formed
during crystallization of a gel. The writer has observed metacolloidal-granular shell texture in sphalerite pisolites of the Iokun'zh
deposit (Fig. 27) and in cassiterite oolites of the Shakh-Shagaila
deposit (Fig. 129).

Allotriomorphic-granular shell texture is apparently secondary in
most cases and is formed by recrystallization of collomorphic-
globular, metacolloidal-granular, and spherulitic aggregates.

In conclusion, we should note that in some cases the character
of the textures of mineral aggregates is the same for all the shells
of an oolite, whereas in other cases there is a progressive change in
texture from the outer shells toward the center, and in still other

cases there is an alternation of shells with different kinds of textures. The galenite–sphalerite pisolites of Iokun'zh, in which there is an alternation of metacolloidal sphalerite shells with characteristic globular–collomorphic texture and galenite shells forming coarsely columnar aggregates (Fig. 26), is a typical example of such an alternation.

Thus, in some cases the textures of the oolite shells settle unambiguously whether the oolites are of colloidal or crystallization origin. In other cases, when the aggregates forming the shells are recrystallized, in order to determine the origin of the oolites either relicts of the the original shell texture must be found or additional criteria (morphological features, etc.) must be used.

A number of papers (Rozhkova and Solov'ev, 1937; Link, 1903; Schade, 1909; Bradley, 1929; Morse, Donnay, and Ott, 1933; Reark, 1952; and others) have been devoted to the experimental production of oolites. It would hardly be advisable to give an exposition of most of this work, because the main results have been reviewed by Chukhrov (1955), but one of the papers, by Rozhkova and Solov'ev (1937), should be discussed. This paper presents the results of many experiments on the conditions of the formation of oolites composed of carbonates of Ca, Mg, Mn, Fe, Cu, and other elements. It is especially interesting that the oolites in these experiments were obtained from amorphous precipitates of these compounds. These were obtained by precipitating calcium carbonate (sometimes Ca and Mg and others) by pouring a half-molar calcium chloride solution into one-molar sodium carbonate solution. The bulky, amorphous precipitate thus

Fig. 28. Experimentally produced calcium carbonate oolites (x 20; after E. V. Rozhkova).

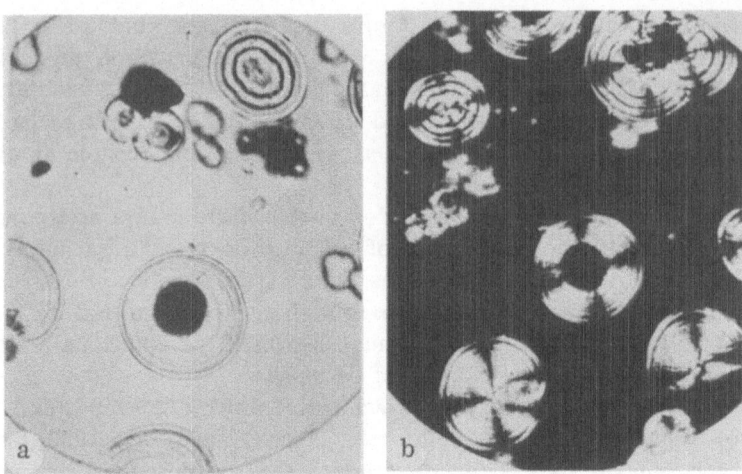

Fig. 29. Experimentally produced calcium carbonate oolites with colored shells
(after E. V. Rozhkova). (a) Uncrossed nicols; (b) crossed nicols.

formed showed distinct globular texture. After some time the gel
decreased sharply in volume, became thicker, and consisted of
oolites and spherulites (Fig. 28). For calcium carbonate the period
of compaction of the precipitate was two or three days; for carbonates
precipitated from mixtures of Mg and Ca chlorides, the time for
compaction of the precipitate and formation of oolites was not more
than four hours.

Rozhkova (Rozhkova and Solov'ev, 1937, p. 360) made experiments
on the continued growth of previously obtained oolites. The oolites
were colored black and then put into a calcium chloride solution to
which was immediately added a sodium carbonate solution in a quantity
about 65% of what was necessary to precipitate the calcium chloride
completely, and "after one or two days it could be seen under a
microscope that the surface of the completely black oolites or their
aggregates showed one or several zones of noncolored material.
The shapes of the oolites or their aggregates around which deposition
of the new material took place determine the shapes of the growing
oolites" (Fig. 29).

Further, Rozhkova noted, "Very often individual oolites grew
together. In our opinion this can be explained in the following way.
During deposition of new material on the surfaces of the oolites, two
or more closely spaced oolites can come into contact. This contact
is between outer, fresh, still semicolloidal surfaces, which, by merging together, form a common zone that is completely homogeneous.

Such a group of oolites, consisting of two or more centers of oolite formation, is itself subsequently the center around which new material is deposited, exactly as described above" (Fig. 30).

SPHERULITES

Spherulites are spherical mineral aggregates with radiating structure.

The structural and textural features and processes of the formation of spherulites have been discussed in detail by Grigor'ev (1953, 1961), who noted that three methods of formation of spherulites are now known.

1. The first method of formation of spherulites is related to the growth of crystals on the surfaces of foreign equidimensional particles.   There is geometrical selection during growth of variously oriented nuclei, with the ultimate result that growth continues only in radially oriented crystals.   A spherical aggregate with radiating structure is thereby formed (Fig. 31a).

2. The second method is similar to the first in the mechanism of formation of the spherulite and differs only in that the center of crystallization is not a foreign particle but an aggregate of variously oriented nuclei of the same material concentrated at individual points (Fig. 31b).

3. The third method is connected with the cleavage of crystals. The causes of cleavage are various.   Cleavage connected with non-

Fig. 30. Asymmetrical calcium carbonate oolite formed by the mantling of an aggregate of small oolites (x 180; after E. V. Rozhkova).

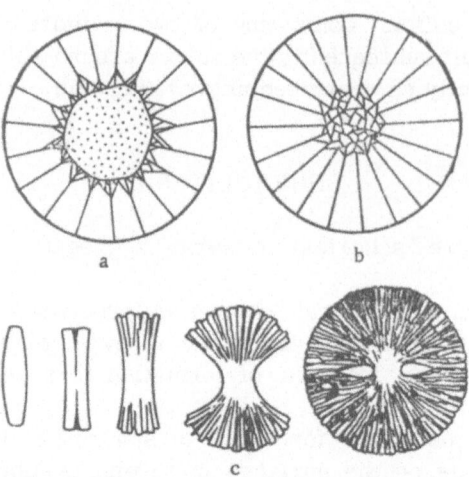

Fig. 31. Diagram illustrating the formation of
spherulites (after D. P. Grigor'ev). (a) From
many nuclei growing on a foreign particle; (b) from
an aggregate of nuclei; (c) from a split crystal.

uniform (sectorial) enrichment of isomorphic or mechanical admixtures has been studied most definitively. In the ideal development
of the cleavage process, a spherulite (spherocrystal) is formed
(Fig. 31c).

There is no need to discuss examples of spherulites formed by
the first two methods, for they are widespread and well known.
Spherulites whose formation is related to the cleavage of crystals
are less common. The writer has observed spherulites of this kind
in calcite in the Upper Carboniferous rocks of the Moscow region
(Fig. 32) and in barite of the Irba fluorite deposit, Krasnoyarsk
Territory (Fig. 33).

It is noteworthy that spherulites are formed not only by direct
crystallization from true solutions but also in substantial numbers
by recrystallization of metacolloidal aggregates. Of considerable
interest in this respect are the observations of Rozhkova, who established experimentally that over relatively short periods of time globular and oolitic forms of calcium carbonate progressively recrystallized into spherulites (Fig. 34) and rhombohedral crystals.

Rozhkova (Rozhkova and Solov'ev, 1937, p. 352) noted the following
stages of alteration of a calcium carbonate gel after it is obtained:
"The first stage is always the precipitation of amorphous calcium
carbonate (perhaps the basic carbonate). The next stage is the transformation of the amorphous carbonate precipitate into a spherulitic

precipitate by physicochemical interaction of the amorphous carbonate precipitate with calcium carbonates and chlorides that are present in the solution both as colloids and as electrolytes. The third stage is the change of the spherulitic precipitate into a crystalline precipitate, caused by the same interactions of the precipitate with the medium."

Rozhkova established that the spherulites are stable at a pH greater than 7.4 to 7.6, whereas "at pH less than 7.4 the original calcium carbonate spherulites recrystallize rapidly to form rhombohedral crystals."

Terziev (1962) cited no less interesting data on recrystallization of oolites into spherulites. In studying the morphology of metacolloidal pyrite from the lead–zinc deposit of the Madan region (Rhodope), Terziev observed spherical forms of pyrite that belonged to a single genetic type but often differed from one another texturally. According to Terziev, oolites with distinct concentrically zoned structure are primary (Fig. 35a). Zoning is caused by alternation of zones of yellowish fine-grained pyrite and black powdery pyrite. In the inner zones there is ill-defined finely columnar texture. In addition to oolites of this type there are oolites in which all the shells have well-defined columnar texture (Fig. 35b). In this case the columnar texture is undoubtedly secondary, the result of recrystallization of

Fig. 32. Calcite spherulites formed by the splitting of rhombohedral calcite crystals. Shchelkovo deposit (× 5).

Fig. 33. Part of a barite spherulite forming by the splitting of a
crystal. Irba deposit (natural size).

metacolloidal-granular aggregates. Oolites of this sort are the most
widespread, and texturally they can be considered transitional to
spherulites. Finally, a third type of spherical metacolloidal pyrite
has distinct radiating structure (Fig. 35c). Considering the absence
of zoning, aggregates of this kind should by textural features be
classed as typical spherulites.

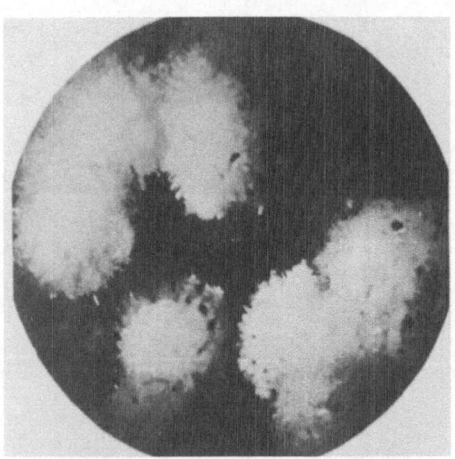

Fig. 34. Spherulites forming by crystallization
of an amorphous magnesium carbonate precipitate seven days old (x 8; after E. V.
Rozhkova).

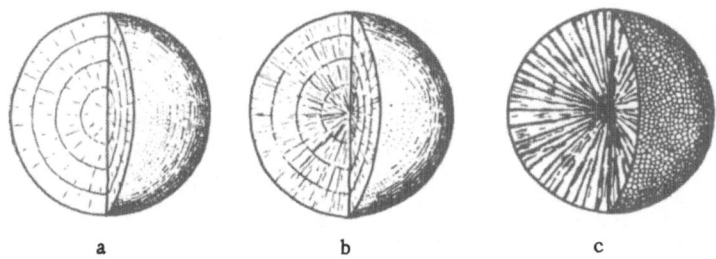

a   b   c

Fig. 35. Diagram illustrating the formation of spherulites by progressive recrystallization of oolites (after G. I. Terziev). See explanation in the text.

Terziev also traced the subsequent stages of recrystallization of the spherical aggregates of metacolloidal pyrite all the way to the formation of distinctive combinations of radially oriented lamellar pyrite crystals and octahedral pyrite crystals.

In studying the features of modern pisolitic aragonite aggregates at the outlet of the Ca–Na–Mg hydrocarbonate waters of the Akhaltsikhe well, the present writer found that, after these aggregates formed (in 1959), they underwent repeated crystallization over very short periods of time (three to three and a half years).

Pisolites one to one and a half years old that fill the pools surrounding the spring and move about freely in them are no larger than 0.5 to 1 cm. Well–defined, thin, concentric parting and distinct zoning are characteristic (Fig. 36a). The texture of the shells is spherulitic. There is evidence of recrystallization: Groups of shells are divided into individual segments with radiating structure. Because recrystallization is incomplete, pisolites of this type retain their primary textures.

After two years the pisolites mentioned above from the same pools grew to 1.5 to 2 cm and some to 3 to 3.5 cm. The pools were filled with pisolites, which were cemented by a tufalike mass of aragonite and in a number of cases had their surface covered by a dense crust of aragonite sinter. When these pisolites were split open, it was found that they had almost completely lost their primary concentrically zoned structure. Most regions of the aggregates showed well–defined radiating structure. Only occasionally were small aragonite relicts showing concentric zoning preserved in the overall recrystallized mass.

Usually recrystallization begins at the centers of the pisolites. At first a spherulite is formed there with nonuniformly developed growth sectors (Fig. 36b). Gradually most or all of the pisolite develops radiating structure (Fig. 36c). Thus, spherulites can also

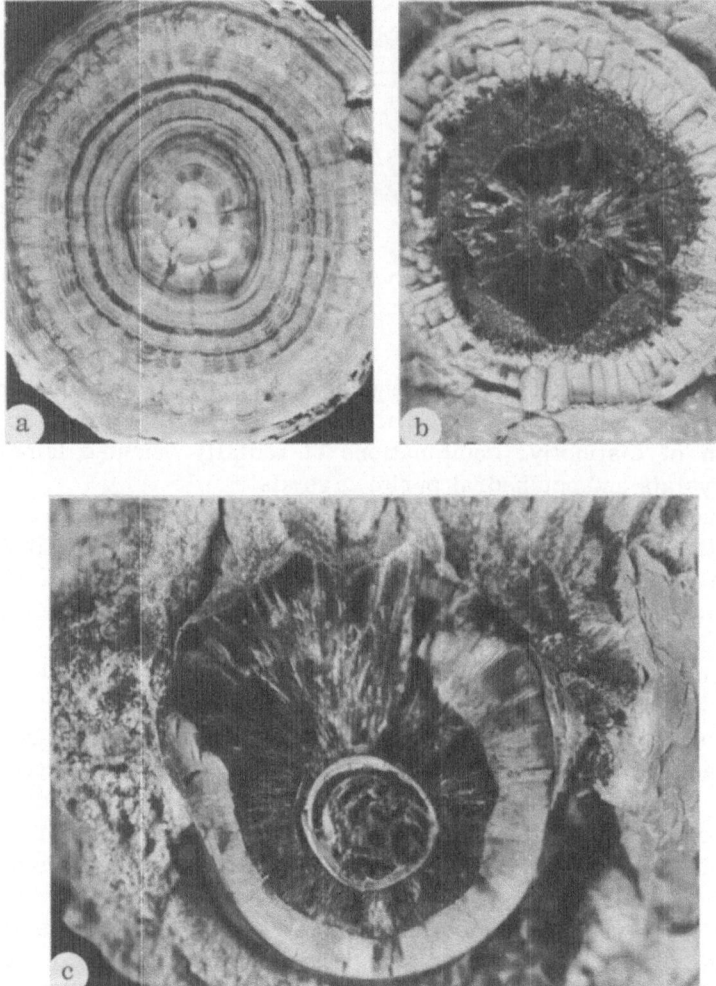

Fig. 36. Recrystallization of aragonite oolites. (a) x 8; (b) x 3; (c) x 2.5.

be formed by recrystallization of spheroidal aggregates of some other type.

In conclusion we emphasize the following: Though spherulites are typically crystallization aggregates, they are not evidence that the crystallization was from true solutions. A crystallization nature of spherulites satisfies the mechanism of formation of these aggregates but does not necessarily reflect the physicochemical nature of the medium.

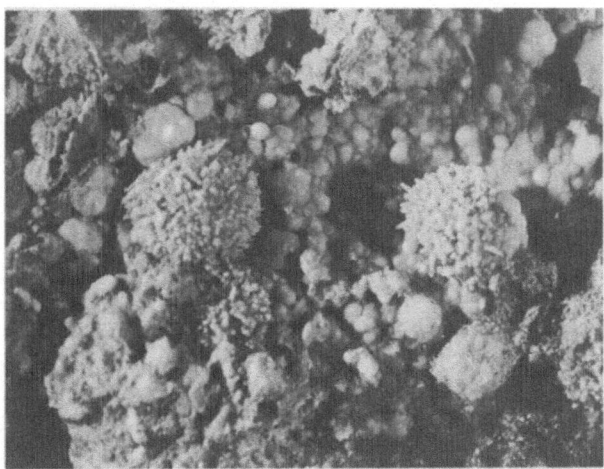

Fig. 37. Spherical membrane forms of cristobalite in fissure cavities of chloritized andesite tuff. Pamach deposit (x 5).

To some extent the question may be raised as to which media are more or less favorable to the formation of spherulites. In this respect, true solutions are undoubtedly among the most unfavorable media. On the other hand, melts and gels are in most respects the most favorable media for the formation and growth of spherulites (volume distribution of accumulations of nuclei, possibilities for the growing spherulite to remain in suspension for a long time, etc.).

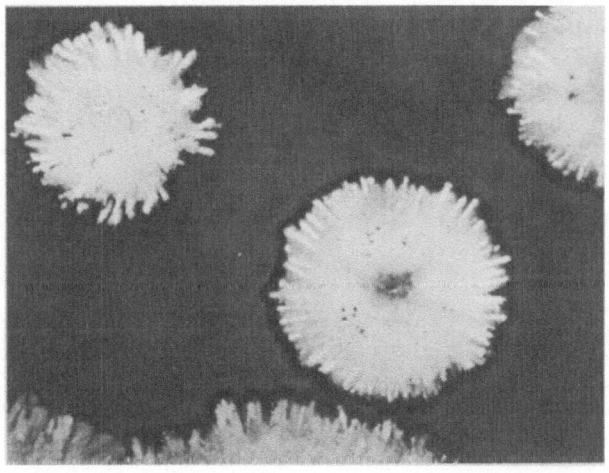

Fig. 38. Spherical aggregates consisting of radial membrane tubes of cristobalite (in chalcedony). Pamach deposit (x 10).

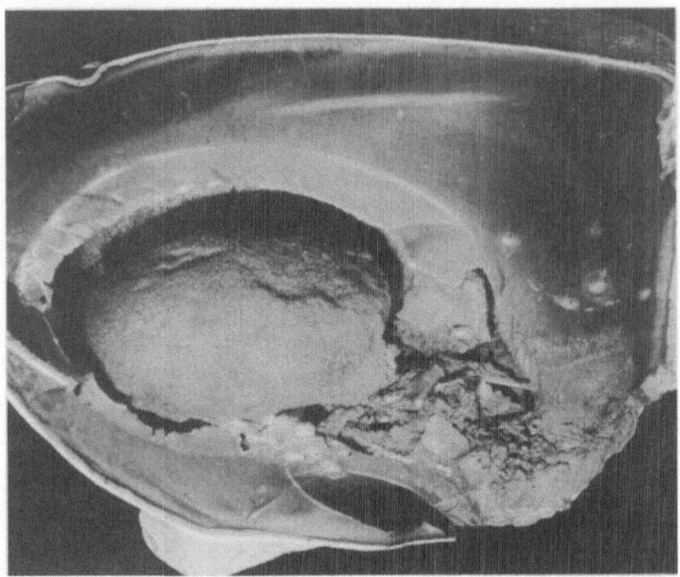

Fig. 39. Exfoliation spheroid in a siliceous concretion from the Carboniferous limestones of the Moscow region (natural size).

Fig. 40. Exfoliation spheroid in a mass of metacolloidal laumontite.
Pauzhetka deposit (x 2).

MEMBRANE-SPHERICAL MINERAL AGGREGATES

Besides the principal and most widespread mineral aggregates described above, the writer has observed distinctive spherical and hemispherical forms composed of radially oriented membrane fila- ments of cristobalite. The membrane-spherical aggregates of cristobalite are abundant in chalcedony–cristobalite and cristobalite– mordenite veins of the Pamach deposit (Georgian SSR).

In these veins there are numerous cavities lined with coarsely glob- ular cristobalite aggregates as thin (0.5 to 2 mm) incrustations. Spher- ical forms of cristobalite with membrane-fibrous structure are un- evenly distributed on the surfaces of these aggregates (Fig. 37). Their size varies from 2 to 7 mm, and they consist of a dense aggregate of radially oriented filaments. Each filament is a tube with a rounded end. The membrane tubes are 0.05 to 0.07 mm across, in some cases reaching 0.1 mm. By growing together, the membrane spheroids form aggregates that are closely similar to reniform aggregates (Fig. 38).

Radial growth of the membrane tubes is caused by the accumulation of a great number of membrane nuclei into certain centers. It is hypothesized that large globules of amorphous silica act as nuclei in this case. Study of silica gels from the Pauzhetka hot springs supports this hypothesis to some extent.

In studying viscous gels under an electron microscope, the writer observed numerous irregular rounded forms with one or two small protuberances. The irregular rounded masses form during coagula- tion of the gel, among whose partially combined aggregates tubular extensions are occasionally observed (see Fig. 63).

We should note that in the main the mechanism of formation of natural membranes is unclear and is in need of the most serious future study.

SPHERICAL MINERAL AGGREGATES OF CONTRACTION TYPE

In the late stage of diagenesis of a gel, during its overall compac- tion, there is quite a substantial decrease in the volume of the gel precipitate. This volume contraction, acting nonuniformly in different parts of the gel, leads to the formation of contraction cracks, which can form concentric fracture systems and produce exfoliation spher- oids if the gel is layered. Chukhrov (1955) noted such phenomena in chrysocolla and azurite from Kazakhstan, siderite from the Komi ASSR, and siliceous concretions from California. The writer has observed exfoliation spheroids caused by contraction in siliceous concretions from the Carboniferous limestones of the Moscow region and in irregular segregations of metacolloidal laumontite in hydrothermally altered tuffs at Pauzhetka (Kamchatka).

Morphologically, exfoliation spheroids in siliceous concretions generally inherit the shape of the primary diffusional zoning of the siliceous gel (Fig. 39). The surface of the spheroid, and also the outer surface of the contraction crack, is very thinly coated with well-crystallized quartz. This indicates that the contraction crack was at some time filled with a solution that had been squeezed into it.

Exfoliation spheroids of metacolloidal laumontite also inherit the elements of diffusional zoning in gels of laumontite composition. Morphologically they are usually irregularly spheroidal segregations in a dense mass of metacolloidal laumontite (Fig. 40); ellipsoidal segregations and ideally formed spheres are less common. They are set off from the main mass by very fine contraction cracks. The surfaces of the exfoliated spheroids of metacolloidal laumontite show signs of crystallization.

It should be noted that among metacolloidal aggregates of various mineral species there are forms that are closely similar to those discussed above but slightly different. The distinguishing feature of such forms is that the exfoliation spheroids are not individualized, because the contraction cracks surrounding them are completely filled with coarse-grained aggregates of the same mineral.

The widespread forms of the contraction type require further study to settle definitely the problem of their relationship to contraction phenomena in gels.

*   *   *

From the foregoing, six types of spherical mineral aggregates can be definitely recognized at present: globules, globulites, oolites, spherulites, membrane spheroids, and exfoliation spheroids. Some of these are exclusively colloidal in origin and are caused by rigorously determined diagenetic processes in gels. Globules, globulites, membrane spheroids, and exfoliation spheroids belong in this category. Others can be formed both by coagulation and by crystallization. Oolites, whose morphology and shell textures usually permit an unambiguous determination of their origin, belong in this category. Finally, spherulites, which form both by crystallization from true solutions and by crystallization of a gel, are aggregates of primarily crystallization origin. Spherulites formed by recrystallization of spherical mineral aggregates of other types are also common.

Chapter 3

Genetic Types of Reniform Aggregates

In descriptive mineralogy, distinctive mineral aggregates with characteristic hemispherically uneven surfaces have long been termed reniform aggregates.

For a long time reniform aggregates have been considered meta-colloidal. Recently this viewpoint has been sharply criticized by Grigor'ev (1961), who proposed a crystallization origin for these aggregates, placing them in the flow-deposited category. Grigor'ev (1961, p. 248) notes: "...Genetically these aggregates are at present viewed in different ways in mineralogy: either simply as precipitates from flowing solutions (flow-deposited) or as the result of crystallization of gels (metacolloidal deposits)."

Study of reniform aggregates observed by the writer, both in ancient endogenic and sedimentary deposits and in modern hot-spring deposits in Kamchatka and the Trans-Baikal, has led to the recognition of a number of different modes of formation of these aggregates. Study of natural and experimentally obtained gels of various compositions has shown that under definite conditions of diagenesis these gels form well-defined reniform aggregates. Direct study of the formation of mineral aggregates at the outlets of hot-spring waters has made possible the distinguishing of reniform aggregates of crust type, which translate the relief of the surface that is being covered. Finally, a number of data show that these aggregates form also by group growth of spherulites. Thus, at present it seems possible to distinguish three genetic types of reniform aggregates.

COAGULATIVE–DIAGENETIC RENIFORM AGGREGATES

During diagenetic changes in coagulated gels, reniform aggregates form at various stages and by various diagenetic processes.

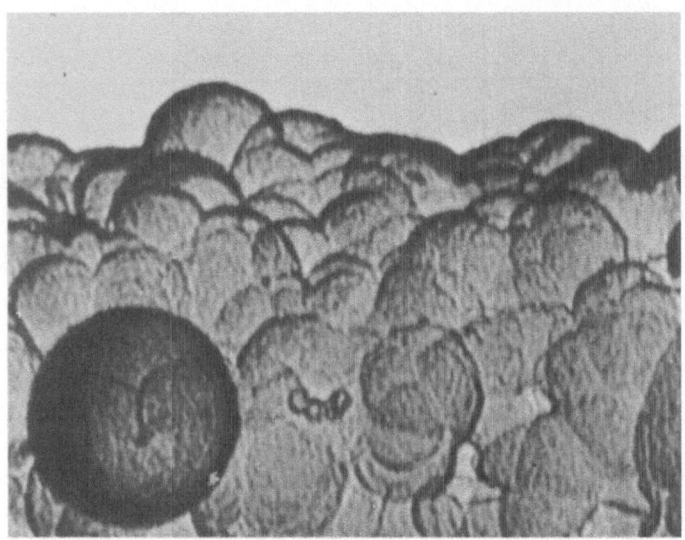

Fig. 41. Microreniform forms of silica gel of coagulative diagenetic type.
Carbon replica (x 80,000).

Study of the natural silica gels that are deposited in abundance at the Pauzhetka hot springs has established that globular aggregates of the gel are often transformed into microreniform aggregates even in the initial stage of diagenesis.

In the globulated gel at some time after its coagulation, there is rather strong coalescence of the globules, with the gel thereby developing a more and more coarsely globular texture. As the gel thickens, the continuing process of coalescence is gradually damped out. The main mass of the gel is then an aggregate of semicoalescent globules, which on the whole has a typically reniform appearance (Fig. 41). The writer has also observed such microreniform aggregates that form during coalescence of experimentally obtained zinc sulfide gels (see Fig. 152).

Upon considerable compaction of the gel, contraction cracks and cavities form, which are generally filled by solutions. As experimental data show (see Chapter 9), the initially even surfaces of these cavities become reniform relatively rapidly (from several hours to two or three days).

The formation of reniform surfaces in the gel is caused by surface-tension phenomena. It should be noted that such surface forms of the gel are observed not only in contraction cracks but also on the surfaces of a gel precipitate that is in a solution of some elec-

trolyte.   Reniform aggregates formed by surface tension are generally macroforms.   In experimentally obtained zinc sulfide gels the
maximum size of the hemispheroidal elements in aggregates of this
kind is 7 to 8 mm in diameter.

RENIFORM AGGREGATES OF CRUST TYPE

The widespread rhythmicity of mineral formation processes causes
periodic, layered deposition of mineral material under very diverse
conditions.   Crusts are often thus formed which clearly translate
the morphology of the coated surfaces. Deposition of material on the
surface of a boulder, on oolitic accumulations, etc., leads ultimately
to the formation of reniform aggregates.

Periodic deposition of material takes place both by rhythmic
coagulation of a sol and by the formation of various sorts of flow-
deposited aggregates.   Thus, depending upon the conditions of genesis,
two types of reniform crust aggregates can be distinguished: (a) coagulative, and (b) flow-deposited.

The reniform aggregates of metacolloidal sphalerite of the Iokun'zh
deposit are examples of reniform aggregates of crust type formed

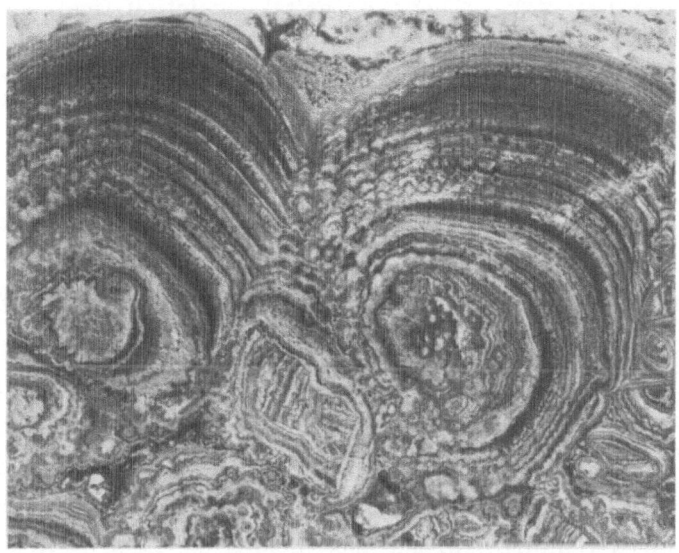

Fig. 42. Reniform aggregate of metacolloidal sphalerite in cross
section. Iokun'zh deposit (x 2).

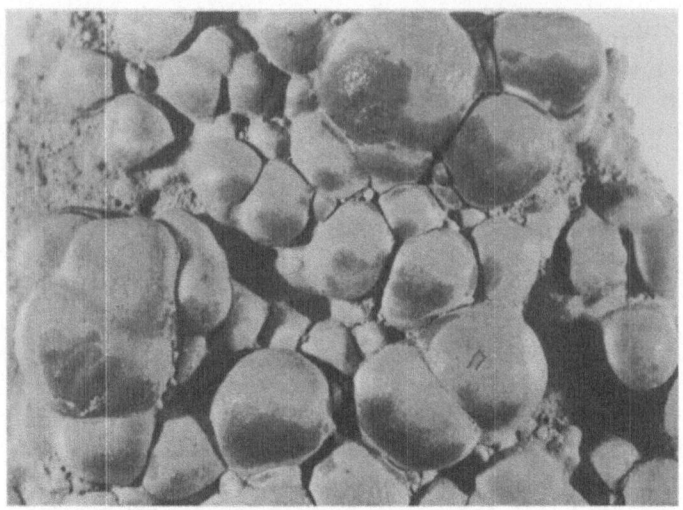

Fig. 43. Reniform aggregate of aragonite formed by the mantling of
pisolites of the same composition by an overall crust.

by repeated coagulative deposition of material on the surfaces of
pisolitic aggregates. The thinly layered sphalerite deposits that
cover the pisolitic aggregates repeat the surface morphology of these
aggregates (Fig. 42). These deposits consist of finely dispersed
metacolloidal–granular sphalerite, containing relicts of the primary
globular–collomorphic texture. There are no signs of radiating
structure either in the pisolites or in the thinly layered crusts
covering them.

A similar form of modern reniform aggregate was observed
by the writer at the Akhaltsikhe well. Over one and a half years the
small pools were almost completely filled by aragonite pisolites. The
surface of these pisolite accumulations began to be covered by an
overall crust of cryptocrystalline aragonite. Growing layer by layer,
the crust, by translating the nature of the surface of the pisolitic aggregate, took on a characteristic reniform aspect (Fig. 43).

Reniform aragonite crusts being formed by outflow of bicarbonate
waters at the Akhaltsikhe well are an example of aggregates of this
type formed from flowing solutions. The mineral waters, ejected
from the well in pulses, flow out over a dam and across shore gravel
into a river. In three years a rather thick (up to 15 cm) aragonite
crust with a reniform surface has formed on the gravel. The size
of the individual reniform elements is 10 cm and sometimes more.
In all cases the surfaces of the developing reniform aggregates quite

clearly translate the configuration of the gravel surface. A characteristic feature of reniform aggregates deposited from flowing
solutions is a distinctive surface sculpture of the reniform elements,
reflecting (1) deposition of material from individual flows and (2) a
certain asymmetry of the surfaces, corresponding rigorously to the
direction of movement of the solutions. We shall pursue these
matters in more detail in analyzing the morphology of flow-deposited
mineral aggregates proper.

RENIFORM AGGREGATES OF CRYSTALLIZATION TYPE

Grigor'ev (1961) studied in detail the formation of reniform aggregates by crystallization as a result of group growth of spherulites.
According to Grigor'ev, the main factor determining the formation of
reniform aggregates of the crystallization type is geometric selection
during group growth of the spherulites. There are two stages of growth:
first, free growth of spherulites that are formed in one way or another;
second, growth under conditions of mutual contact between the
spherulites. According to Grigor'ev, flat contact surfaces between
spherulites are formed in this latter stage of growth. During joint
growth, spherulites with differently oriented centers are developed
nonuniformly: Some widen, but others, because of geometrical selec-

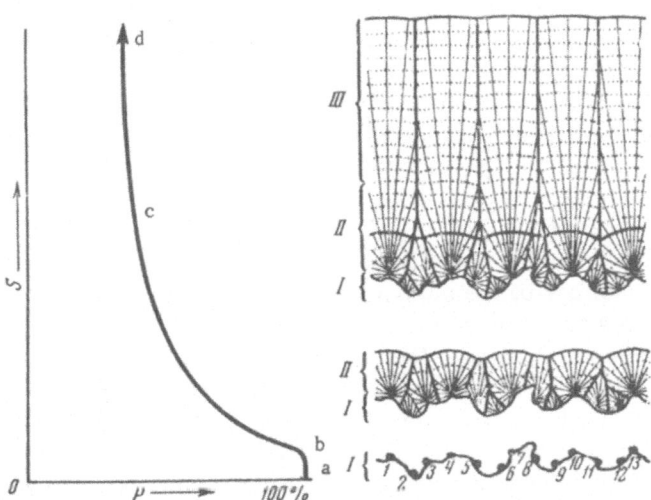

Fig. 44. Diagram illustrating the formation of reniform aggregates
of the crystalline type (after D. P. Grigor'ev).

Fig. 45. Microreniform aggregate of topaz composed of intergrowing
spherulites. Shakh-Shagaila deposit (x 90; uncrossed nicols).

tion, are restricted in growth and ultimately cease to grow. As a
result, certain groups of spherulites "grow out as large segregations
and form the ultimate surface of the aggregate" (Fig. 44; Grigor'ev,
1961, p. 257).

The reniform aggregates of the crystallization type that have
been observed by the writer, which are composed of cassiterite, topaz,
calcite, and other minerals and which reflect the principal features
of joint growth of spherulites, differ from Grigor'ev's picture in some
details of structure.   The main difference is the absence of flat contact surfaces between the spherulites. In the reniform aggregates of
this type observed by the writer, the spherulites in contact always
mutually interpenetrate each other (Fig. 45).   This is in essence
inevitable, because in natural conditions a uniform supply of dissolved
material to acicular radially oriented crystals is almost impossible.
A certain number of the many crystals always develop to be in a predominant position with respect to supply. In view of this, some space
always remains between the radially oriented crystals, into which
crystals (and groups of crystals) of neighboring spherulites can grow.

There can be a certain uniformity of supply during the growth
of spherulites while suspended.   This is possible only in viscous
media (during crystallization of spherulites in melts or gels).

Thus, the formation of flat contact surfaces between spherulites
is almost impossible during joint crystallization of spherulites from
true solutions.

Another no less important difference is that, in the character of the interrelationships between spherulites, all the reniform aggregates of the crystallization type observed by the writer cease to grow at the stage of contact of the spherulites. In isolated cases there was minor growth of spherulites that had come into contact. The cessation of spherulite growth at this stage is in fact caused by the reniform character of the surface of the aggregates. In the opposite case, as a result of further growth there is parallel fibrous growth of crystals by geometrical selection, leading ultimately to the formation of a flat, slightly bumpy surface of the aggregate.

The foregoing facts illustrate the multiplicity of processes that produce aggregates with reniform surfaces. The textures and structures of these aggregates are correspondingly varied. Globular, globular-collomorphic, and metacolloidal-granular textures are characteristic of coagulative-diagenetic reniform aggregates, just as of oolites of colloidal origin. Spherulitic and finely columnar shell texture is most common in flow-deposited aggregates of crust type, and well-defined concentric parting caused by rhythmic deposition is also characteristic. Finally, spherulitic texture with interpretation of spherulites is characteristic of reniform aggregates of crystallization type.

In conclusion we should make a number of critical comments on some of Grigor'ev's ideas and conclusions on the genesis of reniform aggregates.

1. Without denying that reniform aggregates can form by joint growth of spherulites, we shall nonetheless discuss Grigor'ev's (1961, p. 257, Fig. 192) diagram illustrating group growth of spherulites and his remarks on this diagram (see Fig. 44). The diagram illustrates three geometric stages of the group growth of spherulites (Grigor'ev, 1961, p. 257). On the whole, the diagram should correspond to a cross section of the reniform aggregates. But the line of the surface in this pattern is nearly a straight line, or approaches a straight line as a limiting case, corresponding to a flat surface. In describing the stages of geometrical selection reflected in the diagram, Grigor'ev summed up: "As a final result, from the great number of incipient spherulites an aggregate is formed with a reniform surface composed of the larger spherulites, i.e., an aggregate of the type we are concerned with" (Grigor'ev, 1961, p. 258). Thus, geometrical synthesis leads to a flat surface, but logical synthesis proclaims the form to be reniform.

2. While noting that reniform aggregates can form from melts, true aqueous solutions, and gases, Grigor'ev considers that they are formed also from colloids, "but not by desiccation of gels, but rather

from the saturated true solution that fills the interspaces between the particles in any colloid, i.e., again by direct crystallization." Also, "The colloid is probably the supplier of nuclei, both initial and additional, throughout the growth of flow-deposited aggregates" (Grigor'ev, 1961, p. 262). Here above all should be noted Grigor'ev's obvious error. Not a single worker now maintains or has ever maintained that reniform aggregates form by desiccation of a gel. They form during diagenesis of the gel, and, as experimental data show, before the "time of desiccation" the reniform aggregate that has formed becomes rocklike and in the transition to a dry air medium does not dehydrate. Therefore, by the remark "...but not by desiccation of gels..." Grigor'ev leads the reader into error. Further, it is hardly possible to object to the claim that spherulitic aggregates are formed during crystallization of the gel. It is difficult to imagine a medium more favorable to the growth of spherulites, but how flow-deposited aggregates are formed during crystallization of a gel is quite impossible to conceive.

Without dwelling upon other debatable questions of a general nature, we note that Grigor'ev considered the matter of the formation of reniform aggregates by crystallization extremely one-sidedly and not always consistently.

Chapter 4

Flow-Deposited Forms of Mineral Aggregates

In view of the tendency noted above to unite the various morphological types of mineral aggregates under the collective term "flow-deposited aggregates," we should discuss certain features of the formation of flow-deposited aggregates proper.

One of the most characteristic genetic features of flow-deposited aggregates is that their form necessarily reflects the flow process. In the mineralogical literature there are no disagreements in principle on the question of the origin of flow-deposited aggregates. The essence of the genesis of these aggregates was expressed most succinctly by Grigor'ev (1961, p. 247): "It has long been recognized that flow-deposited forms arise when a mineral is precipitated from a solution flowing along some surface and at the same time evaporating." Thus, the flow determines the morphology of the aggregate.

Because solutions circulating along cracks in rocks are also characterized by certain morphologies (drops, streams, sheets), the question arises whether the various morphological types of flow-deposited aggregates do not inherit the forms of flow of the solutions. The writer's observations and a number of literature data allow a positive answer to this question.

For a solution flowing along some surface, we can distinguish three main modes of movement: drops, streams, and sheets. Within each form, several morphological variations, which do not transgress the limits of the given mode of motion, are possible.

MOVEMENT OF SOLUTIONS AS DROPS AND FLOW-DEPOSITED FORMS ASSOCIATED WITH THIS MOVEMENT

Flow-deposited mineral aggregates of various types are formed depending upon the nature of the movement of the drops of solution.

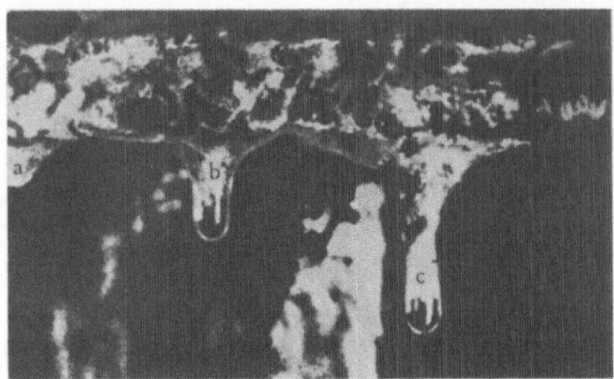

Fig. 46. Different stages of development of present–day stalactites
under the Troitskii Bridge in Leningrad (after A. N. Churakov).

At present, in connection with mineral formation, three types of
movement of drops of solution are clearly evident:

1. Periodic fall of drops under the influence of gravity (dripping)
2. Pressure spraying of drops at some angle to some surface
3. Chaotic movement of very small drops in air

Below is a short description of the flow–deposited mineral aggregates whose formation is connected with these types of motion of
drops of solution.

Flow-Deposited Aggregates Formed in Connection
with Periodic Fall of Drops Under the Influence of Gravity

Drops hanging from a substratum are hemispherical, expanding to
a cone at the base. They are more or less circular in cross section.
Flow–deposited forms of mineral aggregates that are formed by drops
are stalactites and stalagmites. The direction of growth of stalacitites
is the same as the direction of movement of the solution; stalagmites
grow in the opposite direction.

Churakov (1911) directly observed the formation of calcareous
stalactites beneath the arches of the Troitskii Bridge in Leningrad.
The size of the stalactites was determined by the rate of percolation
of the solutions, expressed externally as a greater or lesser frequency of fall of the drops.

Figure 46 is a photograph of stalactites observed by Churakov
in various stages of development. At the left margin of the photograph
is an incipient stalactite in the form of a hemispherical protuberance

with a drop of solution hanging on it (Fig. 46a). The fall frequency of drops of the percolating solution was in this case 15 to 20 min, whereas for the two neighboring stalactites, which have already taken on a cylindrical shape (Fig. 46b and c), the fall frequency of the drops is less than 2 min.

A cross section of the developing stalactites approximates a circle with a diameter equal to the diameter of the drop. The stalactites are hollow in the center, i.e., they are tubes whose walls are composed of concentric crusts of calcium carbonate. Churakov noted that the central cylindrical channel of the stalactites is in places filled with a friable, finely dispersed carbonate mass. Also, hemispherical partitions, convex downward, are always observed in the channels.

Thus, the observational data on direct formation of stalactites show that these aggregates on the whole inherit the morphology of the drops (conical expansion at the base, circular cross section, etc.). Moreover, the direction of movement of the mineral solutions, expressed as the regular orientation of the hemispherical partitions of the central channels, is reflected in the structure of the stalactites.

These regularities are observed not only in modern stalactites, they are also clearly evident in older stalactites. Figure 47 is a photograph of embryonic stalactites from the Carboniferous limestones of the Moscow region. The growth of these stalactites ceased at the very beginning, probably because the solutions ceased to flow.

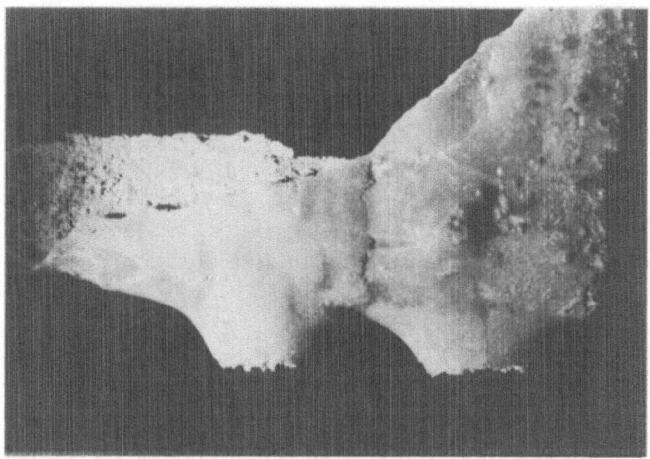

Fig. 47. Embryonic stalactites from sinkholes in Carboniferous limestones of the Moscow region (x 1.5).

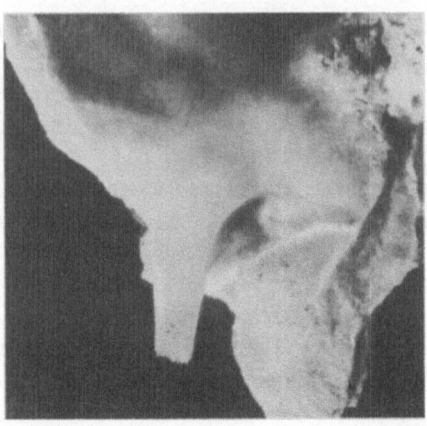

Fig. 48. The most widespread form of stalactites. . Chernorechensk Cavern (natural size).

By comparing his photograph with Churakov's, it is seen that the
cone-shaped embryos of the ancient stalactites do not differ morphologically from the modern ones.

The wide distribution of stalactites and stalagmites is well
known. Cases of flow-deposited aggregates of this type have also
been noted in endogenic deposits (Raibl, Italy; Shurdo, Georgian
SSR).

The stalactites and stalagmites from karst sinkholes in Carboniferous limestones of the Moscow region are of some interest. The
writer has observed them in many limestone caves in the Podol'sk,
Ruz, Vereisk, Serpukhovsk, Ozersk, and other regions.

Of greatest interest are stalactites of nickeliferous calcite of
the Novopodol'sk and Chernorechensk caves. They are small, 1.5
to 15 cm long and 0.3 to 2.5 cm in cross-sectional area. Small,
isolated, funnel-shaped stalactites are most common (Fig. 48). Such
stalactites, interconnected by a thin wavy crust, in many cases
form series of cascading flowstone deposits. Stalactites in the form
of thin, hollow tubes 0.5 to 0.7 cm in diameter and up to 18 cm long
are less common. Sometimes the lower ends of these tubular
stalactites touch limestone projections, grading into dense, pear-
shaped, crusty flow-deposited forms (Fig. 49). The free ends of
stalactites usually are parallel growths of very sharp calcite rhombohedra and less commonly have smooth, enamel-like surfaces.

In cross section the stalactites show concentric structure and
commonly, also, a central tubular cavity sometimes filled with small
calcite crystals. The individual shells are finely columnar calcite
aggregates and are usually sharply set off from one another. The

number of shells is 50 or more, each with a thickness of 0.1 to 2-3 mm. Numerous cases of recrystallization of the finely columnar calcite aggregates can be observed. The aggregates become coarsely columnar, and, because the increase in size of the crystals in one shell is at the expense of adjacent shells, the number of shells in the stalacitites decreases. Transitions of coarsely columnar aggre- gates into homogeneous coarse-grained aggregates, with concurrent loss of concentric shell structure, are slightly less common. In isolated cases stalactites with a single-crystal core in the center, surrounded by a granular fringe, were observed. There is a regular increase in the transparency of the stalactites according to the degree of recrystallization of the aggregates forming it.

The stalagmites have the shape of truncated cones with rounded edges. Their apices are commonly rough and lusterless, but the sides are smooth and enamel-like. A coarse-grained or single-crystal

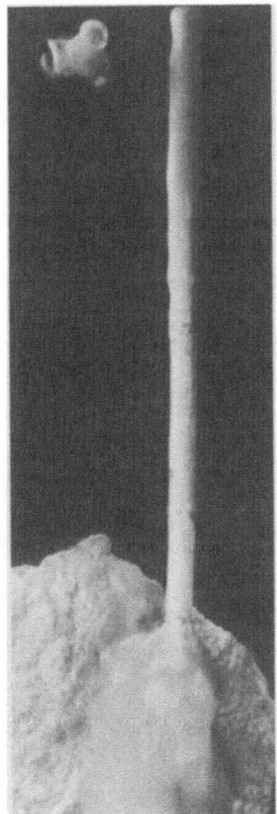

Fig. 49. Tubular stalactite joined at its end with a pear- shaped stalagmitic flowstone. The upper part of the stalag- mite is broken off. Novo- podol'sk Cavern, Moscow re- gion (natural size).

Fig. 50. Formation of a stalagmitic growth
on a tubular stalactite. (a) Original orientation of the stalactite; (b) orientation found
when sample was taken.

central part, often forming more than half of the cross-sectional
area, and an outer zone with concentric structure similar to the
stalactites are observed in the cross sections of the stalagmites.

Very interesting complex flow-deposited aggregates representing
combinations of stalactites and stalagmites were observed in these
deposits. Isolated limestone projections bearing stalactites, in the
karst sinkholes, often break off, and, upon collapsing into cracks,
take on an orientation different from the original. On the surface of
the anomalously oriented stalactite, in areas of intensive irrigation
by drops, a stalagmitic protuberance growing subsequently into a
shape slightly flattened above is formed (Fig. 50). This is a special
type of stalagmite without a flat base.

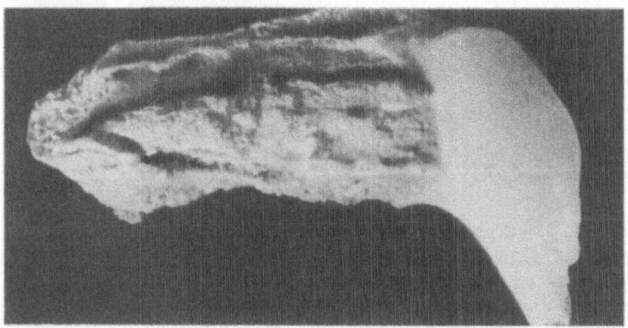

Fig. 51. Formation of an asymmetric stalagmitic growth on the
surface of a broken stalactite.

In another case a phenomenon was noted of a fracture of a funnel-shaped stalagmite and the separation of its two parts without a sharp disturbance of the original orientation. The disturbed surface became a substratum upon which later drops acted, which resulted in the formation of an asymmetric stalagmitic protuberance on it (Fig. 51). The small tubular end of the stalactite also grew asymmetrically.

Thus, anomalies in the shape of stalactites are common. They are generally related not to the mechanism of formation itself but to a change in the geological conditions causing a change in the original orientation of the stalactites and the appearance on them of flowstone surfaces of a different type.

In conclusion, it should again be emphasized that dripstone and flowstone forms (stalactites) related to the movement of drops under the influence of gravity inherit the circular cross section of the drops and the conical shape of their attachment to the substratum and also clearly reflect the direction of movement of the solutions.

Flow-Deposited Aggregates That Form by Pressure Spraying of Drops at an Angle to Some Surface

During outflow of mineralized solutions under pressure, isolated dense jets are accompanied by an aureole of fine discontinuous jets and unidirectionally moving drops. These latter, falling at some angle to the surface of an object surrounding the source, break into a fine spray upon impact. Specific flow-deposited aggregates in the form of arcuate protuberances whose anastomosing systems form a complex cellular crust are thus formed.

Fig. 52. Diagram illustrating the formation of cellular flowstone aggregates. (a) Change in the shape of the drop at the time of impact; (b) structure of the cellular flowstone aggregates in cross section (arrow shows the direction of fall of the drops); (c) structure of these aggregates in plan.

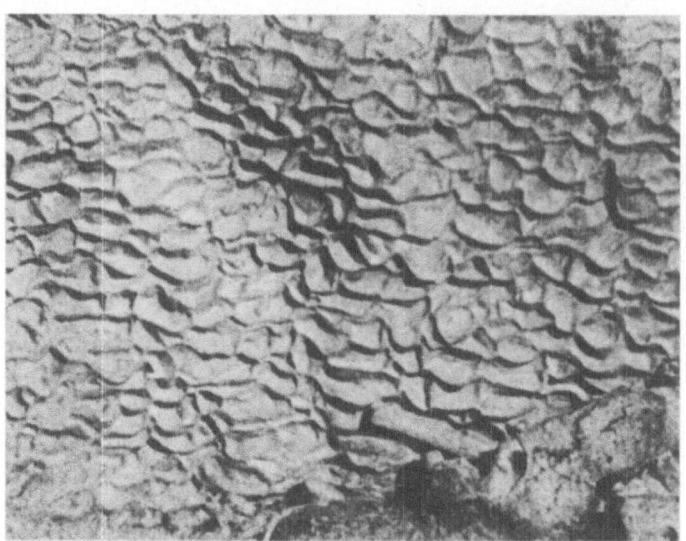

Fig. 53. Cellular crust of aragonite on a boulder around the outlet of
bicarbonate solutions, Akhaltsikhe well (x 2).

The morphology of such aggregates reflects the morphological
changes the drops undergo when they strike the surface of the object.
If we trace the changes the drops undergo at the brief instant of impact upon the solid surface and immediately thereafter, the following
can be noted: At the time of impact the drop flattens out somewhat
and then overflows in the direction of movement, taking on the shape
of a distinctive scale.   In the next instant, it is deflected in the form
of a fan-shaped splash of very fine spray from the surface of the
object at an angle close to the angle of incidence (Fig. 52a). It is precisely in this brief time interval, at the time of contact of the drop
with the solid surface, that mineral material is deposited. The morphology of the flow-deposited aggregates formed in this way to some
extent inherits the scaly shape of the deformed drops (Fig. 52b and c).

The formation of flow-deposited aggregates of aragonite of this
kind observed by the writer at the Akhaltsikhe well is caused by the
liberation of carbon dioxide from solution when the drops strike a
solid surface, which leads to the deposition of calcium carbonate on
the surface (Fig. 53).   Around the outlet of bicarbonate solutions at
this well, cellular aragonite crusts of this kind are formed in a short
time (four to six months) on the surface of various objects (boulders,
boards, cardboard boxes, old shoes, etc.).   The height of the arcuate
protuberances of aragonite reached 1 cm, and sometimes more, in

six months. In the depressions (cells) between systems of anastomosing ridges there are embryonic stalklike forms of aragonite, whose formation will be discussed briefly below.

Chaotic Movement of Drops of Solution in Air
and Flow-Deposited Aggregates Formed as They Settle

When mineralized solutions reach the surface, under certain conditions they are strongly atomized to form aerosols and coarse suspensions of the solution in the air. Upon settling on the surfaces of

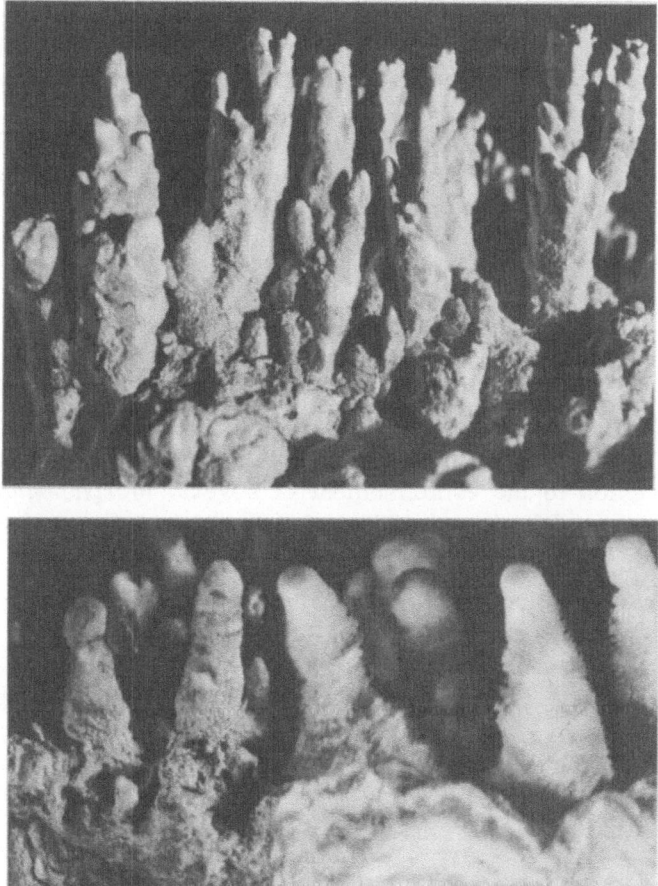

Fig. 54. Stalk forms of sinter around First Paryashchii Spring at Pauzhetka, Kamchatka (x 4).

objects surrounding the outlet, the very small, chaotically moving drops of solution evaporate. Spot-shaped crusts of mineral material remain at the site of the evaporated droplets. Gradually these crusts develop into protuberances and columns. Lateral branches often form on these columns, and the growing aggregates take on the appearance of stalks (Fig. 54). In view of this, the writer proposes that flow-deposited aggregates formed in this way be called "stalk aggregates."

As do stalagmites, stalk aggregates grow in a direction opposite to the movement of the particles of solution. The difference from stalagmites lies in the branching of the aggregates. The aggregates branch because the drops settle not only in the direction of gravity but also along the resultant of two equilibrated forces—the direction of their original complex motion in air and the direction of gravity.

The writer has observed stalk forms of flow-deposited mineral aggregates around the Pauzhetka hot springs (Kamchatka) and at the outlet of bicarbonate waters of the Akhaltsikhe well.

It should be noted that the process of formation of stalk forms of mineral aggregates have as yet been studied very inadequately. They are apparently more diverse and are not restricted to phenomena observed by the writer.

MOVEMENT OF SOLUTIONS AS STREAMS AND FLOW-DEPOSITED FORMS OF MINERAL AGGREGATES THUS FORMED

Observations of modern processes of formation of flow-deposited mineral aggregates from flowing solutions moving as streams and flows have led to the establishment of several morphological types of these flows. Three morphological types of open flows of solutions can be clearly recognized:

1. Planar streams and flows moving along some surface
2. Braided jets that flow vertically off a substratum
3. Streams which run together where they break away from the substratum (as a certain complication of the preceding form).

Flow-Deposited Aggregates Formed by Deposition of Mineral Material from Solutions Moving as Flat Streams and Flows

Streams and flows are a form of movement of mineralized solutions restricted in space to a definite surface, which determines their flat development. The mineral material, precipitated from solution as very thin layers, coats the substratum along which the solutions flow.   The gradual aggradation of material leads to the formation of a crust on the substratum. The morphology of these crusts is to a great extent determined by the nature of the surface of the substratum and the regime of the circulating solutions.

In gradual layer-by-layer growth of material on some object, not only the principal morphological features of the coated objects but sometimes even the finest details of the structure of these surfaces are clearly translated in the surfaces of the aggregates. However, with increasing thickness of the crust these details are translated less and less distinctly. In deposition of material upon a substratum that is more or less even, the crusts also have relatively even surfaces. If the substratum has sharply expressed "microrelief" (lumps of soil, gravel, accumulations of pisolites, etc.), the growing crusts take on a reniform appearance. Some features of the formation and morphology of reniform crusts will be briefly discussed below.

The formation of reniform crusts was observed at the outlet of the bicarbonate solutions of the Akhaltsikhe well. The thermal solutions are discharged in pulses and flow over a dam and across shore gravel into a river as fluctuating streams. By layer-by-layer deposition of material over three years, a thick (up to 12 to 15 cm) aragonite crust with a characteristic reniform appearance was formed on the gravel.   The radius of curvature of the reniform elements corre-

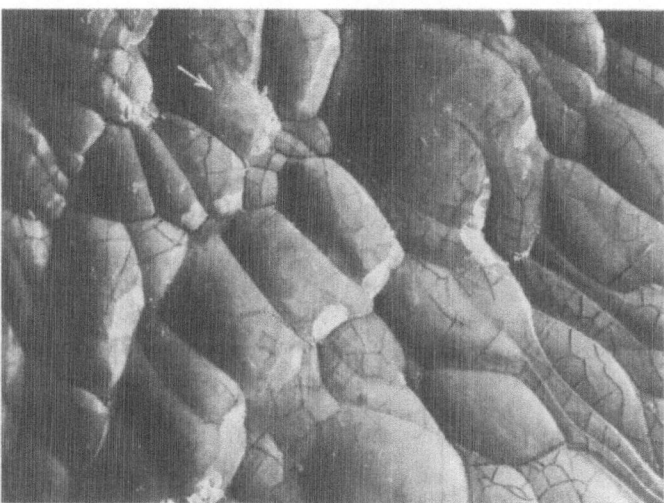

Fig. 55. Asymmetrical imbricated protuberances on the surface of a reniform element of aragonite (arrow shows the direction of flow of solutions). Region around the Akhaltsikhe well (× 10).

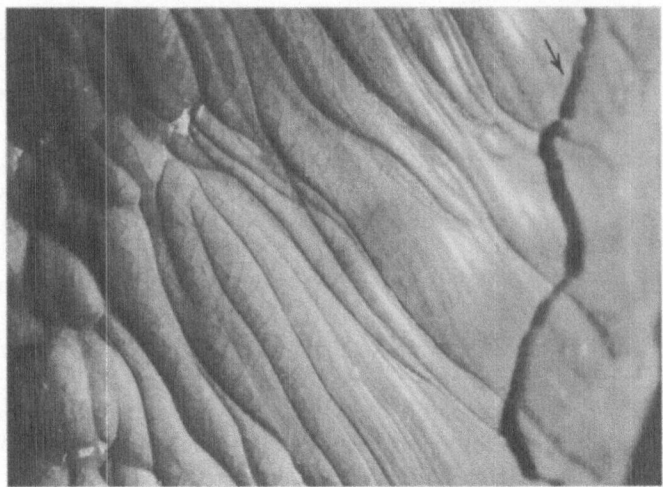

Fig. 56. Striated-uneven sculpture on a lateral surface of a reniform
element (x 2).

sponds to that of the clasts coated by the aragonite. The surfaces of
the reniform elements show distinctive striated-uneven or imbricated
sculpture.

The imbricated elements of relief (Fig. 55) forming the sculpture
of the reniform elements show clear asymmetry of shape in the direction of flow of the solution. The side of the protuberance facing down-
stream is short and steep, and the upstream face is gently spherical
and longer. It should be noted that the imbricated sculpture of the
surface is characteristic only of the upper part of the reniform element; its sides show striated-uneven sculpture (Fig. 56).

These features of structure of the surfaces of the reniform elements are governed by the regime of the circulating solutions. It was
pointed out above that the bicarbonate solutions from the well are
supplied in pulses. Therefore, the flow over the surfaces of the
developing aggregates dries up periodically, breaking up into numerous
thin discontinuous streamlets.

The discharge conditions of the solution are such that for most
of the time the surfaces of the aggregates are washed by these discontinuous separated streamlets. Tracing their behavior on the surfaces of the reniform elements, the following can be observed:
Impinging upon the upper part of the surface, the small streamlets
break up into still smaller streamlets and flow along the sides of the
reniform elements. Flattened, crescent-shaped drops of the solution
remain on the upper parts of the reniform elements. The meniscus-

shaped side of the residual drop matches the width of the waning forward movement of the streamlet and always has the same direction as this movement.   Precipitation of mineral material from the residual drop determines the morphology of the imbricated protuberances.

It is of some interest to compare the surfaces of the reniform elements described above with those formed in pools by periodic layer-by-layer coating of pisolitic accumulations at the bottoms of these pools. Precipitation of material in the latter is slower because of the irregular influx of small portions of solution as drops and separated streamlets.   Also, the mineral material is precipitated in a more tranquil environment, without directed movement of solutions. The elements of the reniform aggregates formed in this way have smooth, enamel-like surfaces.   There are no signs of asymmetric sculpture.

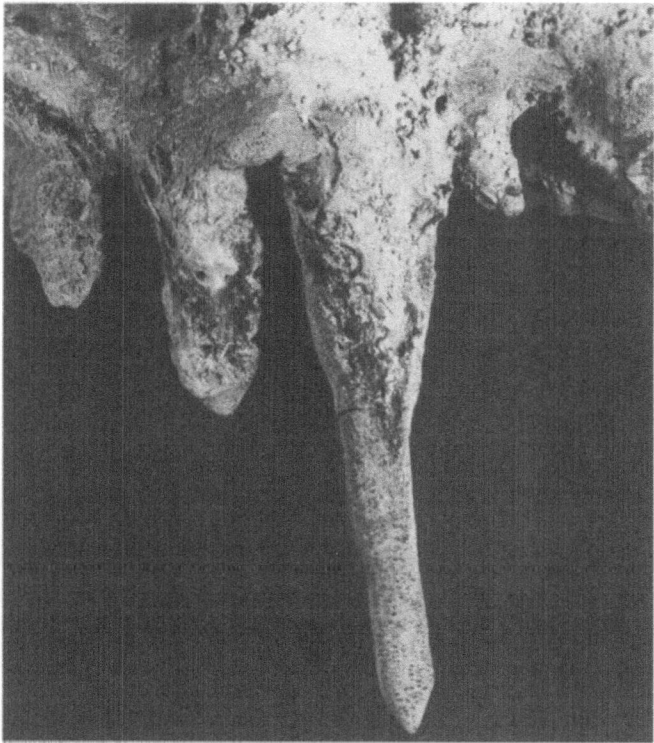

Fig. 57. Stalactitelike flowstone form of aragonite. Akhaltsikhe well
(x ½).

Thus, the presence of specific sculpture on the surfaces of the reniform elements, clearly reflecting the flow direction and to some degree inheriting the specific features of movement of the mineralized solutions, is characteristic of flow-deposited reniform aggregates of the crust type formed by deposition of mineral material from flowing-streams.

Flow-Deposited Aggregates Formed from Braided Jets Flowing Vertically from a Substratum

Flow-deposited mineral aggregates deposited from solutions flowing off a substratum as vertically falling jets are represented by stalactitelike forms. Despite the strong outward resemblance, these flow-deposited forms differ from stalactites: (1) The characteristic conical thickening at the base is absent; (2) the central cylindrical channel is absent; and (3) stalagmites are not formed.

Modern formation of such stalactitelike forms was observed by the writer at the Akhaltsikhe well in the new part of the water-break. Diversion of the water had been going on for six months before the observations were made.

Part of the bicarbonate solutions that are ejected from the well in pulses fall on a reinforced-concrete plate fixed in the discharge. After flowing along its surface they break away from the edge of the

Fig. 58. Comblike deposit of aragonite with characteristic serrated edge. Akhaltsikhe well.

plate as several braided jets. At points of separation of these jets from the substratum there are stalactitelike deposits, ranging in size from 2-3 to 15-18 cm long and from 1 to 3 cm in cross section, on the substratum (Fig. 57). Their surfaces are uneven and bumpy with striated sculpture. Their ends usually are rigorously conical.

In cross section the stalactitelike forms show distinct, concentrically zoned structure caused by layer-by-layer deposition of material. There is no central cylindrical cavity. The concentric shells are composed of very small (0.005 to 0.01 mm) spherical aragonite particles; most of these particles are isotropic, and only isolated ones (not more than 5 to 7%) show signs of radiating structure, in the form of diffuse extinction crosses. In cross section these forms show layer-by-layer conical structure.

Thus, in their cone-in-cone structure these stalactitelike forms deposited from vertically falling jets of solution are more reminiscent of stalagmites. Although in appearance they are conical on the whole, they never show transitions to conical bases and tubular bodies, as is characteristic of stalactites.

Flow-Deposited Aggregates Formed Where Streams Run Together

When mineralized solutions move along the surfaces of such objects as girders and boards, the flowing solutions break up into individual streamlets and run toward the edges of the object. By flowing part of the way around the underside of the object, these streamlets often run together again. Where two streamlets run together a film of solution is formed which connects the lower surface of the object and the merging streamlets. Series of comblike deposits that morphologically inherit the outlines of the meniscus films are formed in this way.

The writer has observed such flow-deposited forms of aragonite on boards and other objects. Direct observations of the formation of such aggregates established the following: In the initial stages of growth, at the point of juncture of the lower surface of the board with the meniscus film an inconspicuous ribbed deposit is formed, which does not extend beyond the film; subsequently this deposit grows in the direction of movement and takes on a flattened, toothed form. Series of such deposits form *en échelon* toothed ridges (Fig. 58) covering the lower parts of the sides of boards.

Thus, in this case the specific forms of flow-deposited aggregates inherit the morphological features of movement of the solutions and reflect the direction of this movement.

MOVEMENT OF SOLUTIONS AS FILMS AND FLOW-DEPOSITED AGGREGATES THUS FORMED

A widespread form of movement of mineralized solutions is as very thin films covering rock surfaces. The features of mineral formation associated with the flow of solutions as films have been studied very little, and our information on the morphology of mineral aggregates formed in this way is even less adequate.

The writer has observed mineral aggregates formed by solutions flowing as films in two cases, at the Akhaltsikhe well and in the Podol'sk deposit. At the Akhaltsikhe well the surfaces of the rein-forced-concrete slabs of the dam on both sides of the main water-break are at all times covered by a thin film of solution, which moves slowly along the surface, a weak influx of solutions thrown out from the well. The flow-deposited forms of aragonite observed on these surfaces are distinctive ribbed forms extending in lines normal to the direction of movement of the solutions. In the direction of movement of the solutions the ribbed deposits form a meniscuslike curvature with convex sides pointing in the flow direction. The deposits are arranged *en échelon* in plan. In the Moscow region (Podol'sk deposit) a similar type of deposit of zinc- and nickel-bearing allophane was observed on the surfaces of limestone blocks in Middle Carboniferous deposits.

The features of formation of the flow-deposited mineral aggregates from solutions flowing as films is in need of serious study.

* * *

To sum up, we should note the great diversity of forms of movement of mineralized solutions, not nearly exhausted by the above examples, and we should emphasize that flow-deposited forms of mineral aggregates formed by precipitation of mineral material from these solutions inherit the flow geometry and reflect, in fine details of structure, the direction of movement of the solutions.

Chapter 5

Dunelike Aggregates

Besides the widespread mineral aggregates described above, particular forms of mineral aggregates are sometimes found which correspond to the morphology of sand dunes and which clearly record the upward movement of hydrothermal solutions. The writer (Lebedev, 1953) has observed such dunelike forms of quartz in the Dzhida deposit in a vertical dipping vein in porphyritic granite.

The principal vein mineral is milky quartz, filling the center of the vein and forming 75 to 80% of the vein. The remaining 20 to 25% consists of orthoclase, distributed near the vein boundaries.

Several empty fissures 1.5 to 2.5 cm wide parallel to the vein contacts are observed in the vein quartz. The walls of these fissures are covered with a crust of drusy quartz with distinctive dune-shaped protuberances consisting of dense hornfelslike quartz around each individual crystal.

The protuberances of hornfelslike quartz are asymmetric in shape, with the gentle slope directed down the dip of the vein. The steep slope of the protuberance is always adjacent to a lower (with respect to the dip of the vein) face of a drusy quartz crystal. In some cases these crystals are completely "buried" and in other cases only two or three of the terminal rhombohedron faces, facing up to the dip of the vein, remain free.

In plan these forms give the impression of a system of sand dunes, which is why the writer calls them dunelike (Fig. 59). On examining them under a binocular microscope it is seen that the surfaces of the dunelike protuberances are crystallized and consist of uniformly developed rhombohedral ends of very small quartz crystals.

In places there is a snow-white kaolinlike mineral in depressions between protuberances. Several pieces of evidence indicate that this mineral is gearksutite: scratching of glass, the great amount of water liberated upon heating in a closed tube, a mean index of refraction of 1.448 ± 0.001, and an X-ray picture obtained in the X-ray

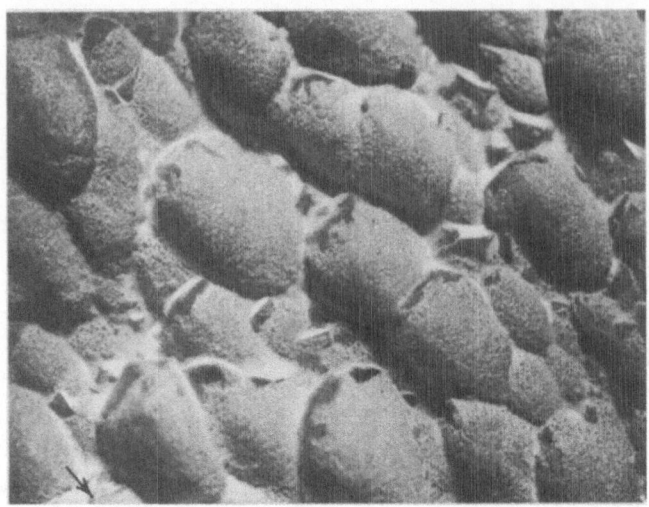

Fig. 59. Dune forms of hornfelslike quartz on drusy quartz (arrow
shows dip direction of the vein) (x 5).

structures laboratory of the All-Union Research Institute of Mineral
Raw Materials. The interrelationships between the drusy quartz and
hornfelslike quartz are clearly observed under the microscope in
oriented sections. The hornfelslike quartz is a dense, finely crystalline aggregate with well-defined granoblastic texture. Toward the
surface of the protuberance the aggregate becomes finer grained. The

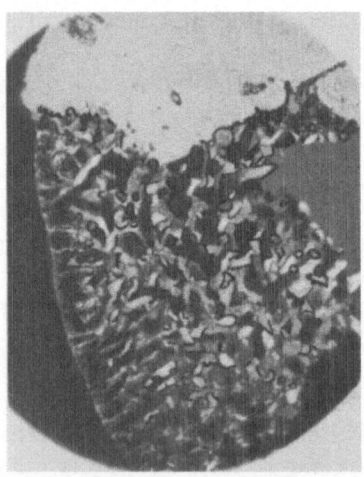

Fig. 60. Corrosion of drusy quartz at
contact with a dunelike protuberance
(x 45).

contacts between these aggregates and the drusy quartz crystals is uneven and corroded, and the faces of the ends of the drusy quartz crystals show very distinct corrosion (Fig. 60).

In the sections studied, regions of hornfelslike quartz are quite common in the comby aggregates of quartz forming the drusy crust. The outlines of these regions are highly diverse and extremely irregular. Increase in grain size from the center of such regions toward the margins and a gradual transition to comby quartz is charac- teristic (Fig. 61).

In connection with the above it should be noted that in the deposit there are a considerable number of veins filled by dense hornfelslike quartz which in external appearance and under the microscope do not differ at all from the quartz aggregates forming the dunelike protu- berances. These veins are found both in the porphyritic granites directly and in the surrounding sedimentary and volcanic sequence. Quartz is represented by dense chalcedonylike and hornfelslike varieties in the parts of the veins near the contacts. Toward the centers of the veins these varieties grade into a quartzitelike quartz that fills the centers of the veins. Lenticular segregations of coarsely crystalline fluorite are observed in this quartz.

Well-formed fluorite crystals, predominantly showing {111} and combinations of {100} and {110} , and formed by free crystallization in the mass of silica gel, are quite uniformly distributed in the hornfels- like varieties.

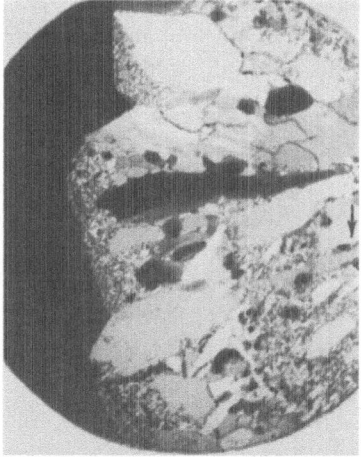

Fig. 61. Crust of drusy quartz with dunelike forms (arrow shows dip direc- tion of the vein) (x 10).

The following picture of the formation of these dunelike aggregates seems likely:

1. Silica was ejected as an aerosol in which the dispersed phase was very small particles of SiO_2 and the medium was probably silicon fluoride.
2. As the aerosol moved upward along the fissure, it reacted with water condensed on the faces of the drusy quartz crystals and was transformed into a hydrosol and a gel. In this, the silicon fluoride, upon being hydrolized, gave silica gel on the one hand and free hydrofluoric acid on the other. The latter reacted with the drusy quartz crystals, partially dissolving them, and the faces directed downward along the dip of the veins were most strongly dissolved.
3. Both the rigorously oriented arrangement of the dunelike protuberances—at the lower faces of the drusy quartz crystals— and their shape are a natural consequence of the character of the transfer of silica as an aerosol.

Accumulation of silica particles around the drusy quartz crystals during transport as aerosols proceeded in the same way that sand dunes form. As the aerosol moved upward along the fissure, the finely dispersed silica particles, on meeting an obstruction in the form of a quartz crystal, were stopped and accumulated around the lower faces of the crystals; hydrolysis of the medium promoted this accumulation, leading to deposition of the entire dispersed phase at places of hydrolysis.

In conclusion it should be noted that Kormilitsyn (1951) has described similar forms of chalcedony on surfaces of comby quartz in one of the deposits of the Durulguevsk massif in the Eastern Trans-Baikal.

Chapter 6

Membrane Types of Mineral Aggregates

The formation of membrane types of mineral aggregates is caused by processes in semipermeable membranes. Traube (1867) made the first detailed study of semipermeable membranes; he considered them to be molecular sieves, impermeable to molecules whose size was larger than the openings between the particles forming the membrane precipitate. By reaction of copper chloride and ferric chloride with K_4FeCN_6, Traube obtained hollow membrane cells whose flexible walls consisted of copper ferricyanide in one case and of Prussian blue in another case. With the cells of Prussian blue, long tubular outgrowths were observed to form. The determining factors in this are the increased osmotic pressure within the membrane cells and the nonuniform thickness of the flexible walls. In the thinnest and thus the most flexible parts there is stronger distortion of the cell walls, with the formation of tubular-fibrous outgrowths.

A wide and extremely valuable survey of work on the theory of semipermeable membranes and the experimental production of osmotic filaments is given in a monograph by Chukhrov (1955). Considering that there is almost no information on membrane types of mineral aggregates in the mineralogical literature, such aggregates will be described briefly below.

The writer has observed membrane types of silica minerals in silica gels deposited in the thermal creeks of Pauzhetka (Kamchatka) and in cristobalite from the Trans-Caucasus agate deposits. In studying the Pauzhetka silica gel (for a more detailed description, see Chapter 10) it was established that in gels with a viscous consistency there are great numbers of membrane tubes, in addition to globular particles, with a maximum size of about 17 to 18 μ long and 0.3 to 0.8 μ in cross section. Irregularly rounded forms of membrane-cell type (Fig. 62) are often observed in addition to tubes. Study under the electron microscope shows that such rounded forms, which are quite a bit larger than globules, were produced by the coalescence

71

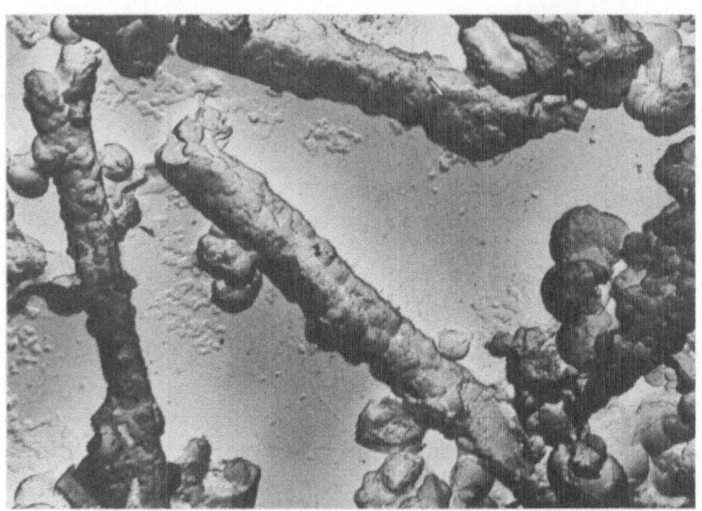

Fig. 62. Membrane tubes and irregularly rounded forms of silica gel
(of the membrane-cell type). Carbon replica (× 24,000).

of globules. This is supported by the presence of transitional types
represented by semicoalescent globules; in a number of cases, it is
precisely from such types that the tubular outgrowths are formed
(Fig. 63).

The writer envisions the following mechanism of formation of
the membrane filaments in the silica gel. The larger, irregularly
spheroidal cells that are formed by the coalescence of globules can
behave like membrane cells under certain conditions. The determining
factor here is the dilution of the solutions containing the gels (and
this is observed periodically for most of the thermal springs and
creeks). Within the spherical shells the concentration of the solution
with respect to its saturation by H_2SiO_3 is a maximum. This concentration corresponds to a definite value of the osmotic pressure.
In the normal condition of the system, the solutions outside and inside
the cell are about equally saturated with H_2SiO_3 (or $HSiO_3^-$). When
the solution containing the gel is diluted, this equilibrium is disrupted,
the osmotic pressure within the spheroidal cells is greater, and tubular
outgrowths—osmotic filaments—begin to form.

It should be noted that these considerations are of course in need
of experimental verification and a more rigorous theoretical groundwork.

Membrane forms are very characteristic of metacolloidal cristobalite, widespread in the agate deposits of the Trans-Caucasus—
Pamach and Shurdo in Georgia and Idzhevan in Armenia.

In the Shurdo deposit, membrane types of cristobalite were observed in chalcedony amygdules in tar-black glassy andesites. Cristobalite as opaline white masses forms marginal fringes in the chalcedony-filled amygdules.   These fringes in contact with the country rock are massive and have a uniform glassy structure. Their inner surface is formed of membrane filaments showing a rough orientation perpendicular to the surfaces bounding the amygdules. Their ends project into the chalcedony mass filling the centers of the amygdules.   In many cases the continuity of the fringes is disturbed, and they break down into isolated, radially membranous spheres. Stalactitelike outgrowths of the marginal fringes, everywhere consisting of membranous cristobalite, are also common (Fig. 64).

Bushy forms of membranous cristobalite precipitated in vesicles were also observed in this deposit in brownish-gray andesites. The size of these bushy segregations is 2-3 mm to 1-1.5 cm long and 1 mm to 0.7-1 cm in diameter. They are fanlike-divergent, branching tubular forms with rounded ends.   Their orientation in the vesicles is highly diverse.   They are observed on the upper walls as well as on the lower and lateral walls, growing vertically and at various angles to the walls.

The size of the membrane tubes of cristobalite varies from 0.5 to 4 mm long and 0.3 to 0.5 mm in outer diameter. Because the tubes are transparent, the thickness of their walls, 0.03 to 0.05 mm, could be measured.   The tubes are often curved, and spheroidal and pear-

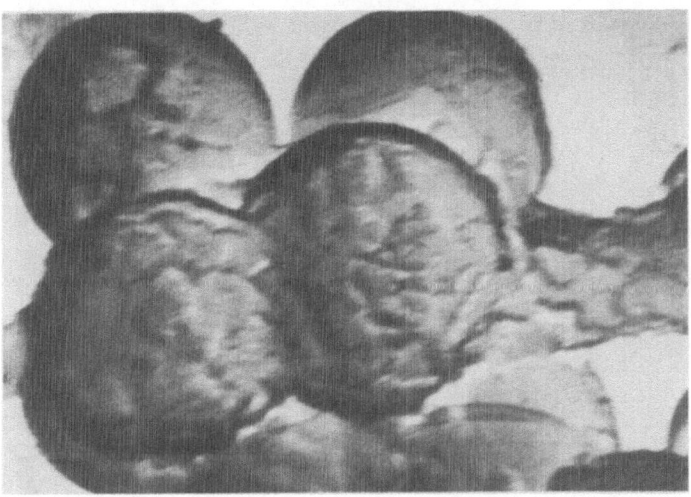

Fig. 63. Semicoalescent globules of silica gel with tubular (membrane) extension. Carbon replica (× 60,000).

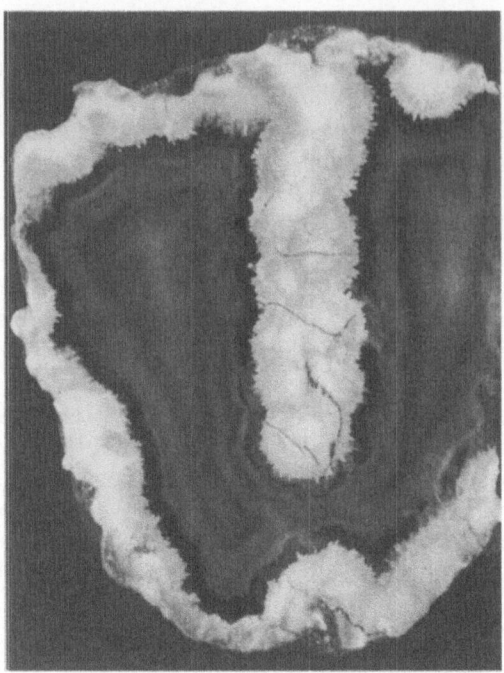

Fig. 64. Stalactitelike form of membranous cristobalite. Shurdo deposit (× 2).

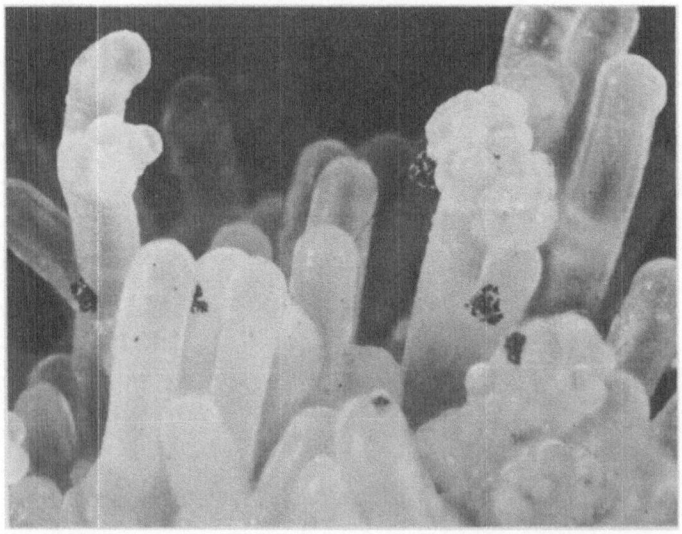

Fig. 65. Membranous forms of cristobalite in a vesicle. Shurdo deposit
(× 30).

shaped accumulations of cristobalite macroglobules are observed on their surfaces (Fig. 65).

The membrane types of cristobalite described above are macroforms of similar membrane tubes in gels of the Pauzhetka deposit.

The distinctive membrane-spherical aggregates of cristobalite of the Pamach deposit were described above.

In the Idzhevan agate deposit, membrane types of cristobalite are observed in amygdules together with chalcedony. It should be noted that metacolloidal cristobalite is extremely widespread in the deposit and together with chalcedony is one of the principal components in the amygdules. They form about 50% of the filling on the average. The cristobalite content varies from 5 to 80% in the various amygdules (Fig. 66). Cristobalite in the amygdules is segregated as fine cristobalite–chalcedony aggregates, in some cases distinctly set off from the uniform mass of pure chalcedony. Gradual transitions from the cristobalite–chalcedony mixture to pure chalcedony are nonetheless common.

Most of the cristobalite in these mixtures is present as membrane types, as accumulations of very fine branching and closely interwoven tubes buried in the chalcedony mass (Fig. 67). It is difficult to determine the length of the tubular filament owing to their complex sinuosity; in any case their length is of the order of centi-

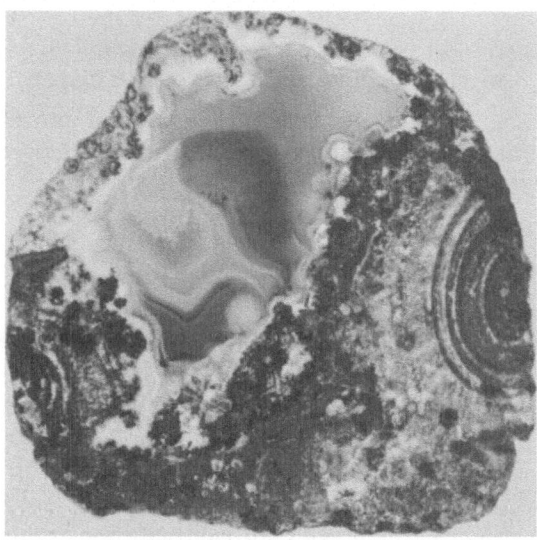

Fig. 66. Structure of a cristobalite–chalcedony amydgdule.
Idzhevan deposit (natural size).

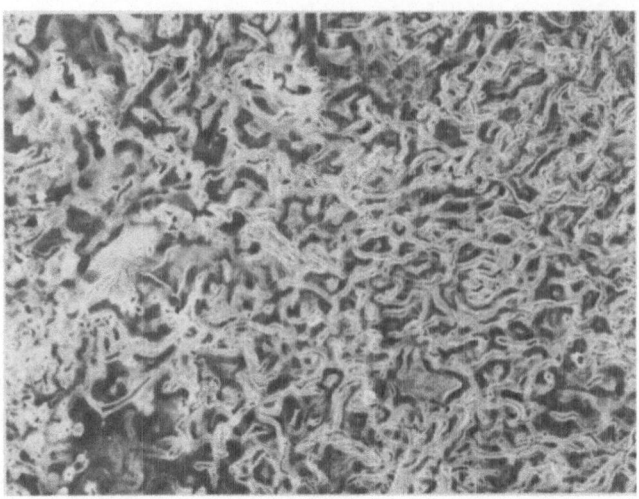

Fig. 67. Accumulation of branching and interweaving membrane tubes of cristobalite in a chalcedony mass.   Idhzevan deposit (x 30).

meters.   The diameter does not exceed $25\,\mu$, and the thickness of the walls is $5\,\mu$ on the average.

Spheroidally lumpy forms of cristobalite with characteristic tubular protuberances reflecting the initial stages of growth of the membrane filaments are also found in the cristobalite–chalcedony mixtures.

Without going into the problems of genesis of cristobalite–amygdules as a whole, we note only that in the metacolloidal silica aggregates, as in silica gels, membrane types of mineral aggregates are not uncommon.   It should also be emphasized that the silica metacolloids studied by the writer generally consist of cristobalite.

BRIEF CONCLUSIONS

Consideration of a number of questions of the genesis of meta-colloidal and superficially similar aggregates shows that their colloidal or crystallization origin is more or less reflected in the aggregates themselves (the features of morphology and texture).

Comparison of the textural features of gels of various compositions but similar consistency (liquid, viscous, dense, etc.) has made it possible to distinguish general textural features common to all the gels at definite stages of diagenesis.

Recognition of specific textures characteristic of all the gels regardless of composition permits us to interpret more nearly unequivocally the textures and metacolloidal mineral aggregates.

As a rule, in specific features of morphology (signs of plastic deformation, contraction phenomena, etc.) and texture (globular, microcollomorphic, spherulitic, etc.) by far the greater part of metacolloidal mineral aggregates correspond to a colloidal nature of the material. It is precisely to this conclusion that comparative study of the texture and morphology of gels and metacolloids leads.

In conclusion it should be noted that flow-deposited aggregates, which are sometimes substituted for the concept of metacolloidal aggregates, always reflect in their structure both the form and the direction of movement of the flows of the solutions from which they were formed and therefore cannot be confused with superficially similar forms of a different type.

PART II

Deposits of Mineral Colloids and Metacolloids

Chapter 7

Mineralogy of the Shakh-Shagaila
Tin Deposit

The Shakh-Shagaila tin deposit is a characteristic example of a deposit intermediate between the cassiterite–quartz and the cassiterite–sulfide type. Smirnov (1947) first recognized tin deposits of intermediate type in the Yano-Adychan region. Levitskii (1939) noted features in common with both cassiterite–quartz and cassiterite–sulfide deposits in the Sherlovogorsk deposit in the Eastern Trans-Baikal. By studying mineral associations, Radkevich (1951) placed a number of Trans-Baikal deposits in the intermediate category. By detailed study of these deposits and analysis of material cited in the literature, Grigor'ev and Dolomanova (1956) recognized tin deposits of intermediate type as a separate cassiterite–quartz–sulfide type. Tin deposits of intermediate type are here noted in Central Kazakhstan for the first time.

The Shakh-Shagaila deposit is located in the northeastern border zone of the Kaib massif, within the Shakh-Shagaila subsidiary granite intrusion. Administratively it is in the Zhanarka rayon of the Karaganda oblast', 97 to 98 km northwest of the Myn-Aral railroad station.

V. I. Volobuev, a geologist of the Kazakhstan Geological Administration, discovered the deposit in 1949. A report by M. Ts. Medoev contains the first geological description of the deposit area; Medoev also made a more or less detailed map of the deposit (in 1951).

The mineralogy of the deposit has received almost no attention, although it is precisely in mineralogical respects that the deposit is most interesting, first, because of our still inadequate knowledge of tin deposits of the intermediate type and, second, because of the relatively wide development of aggregates of metacolloidal cassiterite and quartz in the deposit, attesting to the importance of colloids in endogenic ore deposition.

Fig. 68. Coarse-grained porphyritic
granite (x 30; crossed nicols).

BRIEF GEOLOGICAL DESCRIPTION

The Shakh-Shagaila subsidiary intrusion (Koptev-Dvornikov et al., 1960), which contains the deposit, is one of the largest subsidiary intrusions of the Kaib granitic massif. The intrusion, 7 km east of the Shakh-Shagaila well, is in the border zone of the Kaib massif and in part extends beyond the boundaries of the massif. In plan, the intrusion is oval, with irregular, sinuous boundaries that extend north-south. Occupying an area of about 8 km^2, the intrusion forms a gentle

Fig. 69. Fine-grained aplitic granite
(x 30; crossed nicols).

hill whose surface is mostly bedrock. The relative elevation of the hill is 45 to 50 m. The outline of the hill coincides almost completely with the boundaries of the intrusion. It has been established from borehole data that the intrusion is a stock that dips steeply toward the Kaib massif, with which it is in contact along its southern boundary. To the north and northeast the intrusion is in contact with an Upper Silurian sandstone and shale sequence which in the immediate contact zone has been metamorphosed to biotite-amphibole and biotite-feldspar hornfels.

Within the ore area the granite of the intrusion is of two varieties: coarse, pinkish-red porphyritic granite and fine, pinkish aplitic granite. Coarse, strongly porphyritic, leucocratic granite is the predominant rock type. In the fine-grained groundmass of this rock there are sharply distinct phenocrysts consisting of tabular crystals of potash feldspar, equidimensional grains of dark gray quartz, and, less commonly, thinly tabular crystals of greenish-gray plagioclase (albite no. 5; Fig. 68). The size of the potash feldspar and quartz phenocrysts ranges from 3 to 7-8 mm, and the size of the plagioclase phenocrysts, from 1 to 3 mm. The phenocrysts form up to 70 to 75% of

Table 1. Chemical Composition of Porphyritic Granite

Component	%	Atomic proportions	Number of oxygen ions	Number of cations in "standard cell"
SiO_2	79.70	1328	2656	676
TiO_2	0.08	1	2	0.5
Al_2O_3	11.23	220	330	112
Fe_2O_3	0.33	4	6	2
FeO	0.42	6	6	3
MgO	0.05	3	3	1.5
Mn	Trace	—	—	—
CaO	0.25	5	5	2.5
Na_2O	3.08	100	50	51
K_2O	4.47	96	48	40
F	None	—	—	—
S	0.03	—	—	—
H_2O^+	0.25	34	17	17
Sum . . .	100.23		3140	914.5

"Formula" of rock:

$$K_{49}Na_{51}Ca_{2.5}Mg_{1.5}Fe_3^2Fe_2^3Al_{112}Ti_{0.5}Si_{676}O_{1583}(OH)_{17}$$

the rock. The groundmass, outwardly reminiscent of aplite, is a fine-grained, equigranular aggregate of quartz, potash feldspar, albite, and minor, evenly distributed small biotite flakes. The groundmass penetrates the phenocrysts along cleavages, corroding the crystals slightly.

In the southwestern part of the ore area there is fine-grained, light-pink aplitic granite (Fig. 69), in which the fine-grained groundmass is sharply predominant. Phenocrysts of potash feldspar, albite, and quartz form not more than 20% of the rock, and in some places the proportion of phenocrysts falls to 10%. The mineral composition of this granite is similar to that of the groundmass of the porphyritic granite. There are many masses of aplitic granite extending east-west in the central part of the ore area. The transition between aplitic granite and coarse porphyritic granite is gradual. Both kinds of granite show traces of postmagmatic alteration, expressed as slight alteration of biotite and albite to sericite.

Compared to the average granite, these granites are characterized by extremely acidic composition and almost complete absence of dark minerals. Table 1 gives a chemical analysis of the coarse-grained porphyritic granite, made in the central chemical laboratory of the Institute of Geology of Ore Deposits, Petrography, Mineralogy, and Geochemistry (as were all other chemical analyses cited in this chapter, unless stated otherwise) by O. Ya. Nikolaeva. The analysis was recalculated by the "standard cell" method (Barth, 1955), taking into account the reasonable recommendation of Chetverikov (1956) that the volume of the standard cell be increased tenfold. On comparing the "formula" of this granite with the formula of the average granite (the chemical composition of Daly's average granite was recast) the extremely high silica content, which places this granite in the category of ultra-acidic varieties, becomes evident.

1. The formula of the porphyritic granite

$$K_{49}Na_{51}Ca_{2.5}Mg_{1.5}Fe_3^2Fe_2^3Al_{112}Ti_{0.5}Si_{676}(OH)_{17}O_{1583}$$

2. The formula of the average granite

$$K_{46}Na_{58}Ca_{19}Mg_{11}Mn_8Fe_{10}^2Al_{147}Ti_3Si_{604}P_1(OH)_{46}O_{1554}$$

GENERAL DATA ON STRUCTURE OF THE ORE AREA AND STAGES OF MINERALIZATION

The deposit is restricted to a thick zone of crushing in the central part of the ore area that extends roughly east-west. In the axial part of this zone there are lenses of vein quartz that are interconnected by thin veinlets and are accompanied by a thick aureole of altered rock. North and south of the zone of crushing there are series of *en échelon* latticework zones consisting of mutually intersecting parallel veinlets whose formation was related to the filling of fractures of various sorts. The granite in the latticework zones is altered to quartz—topaz—muscovite greisen.

In the northern part of the zone of crushing, quartz greisens and quartz—topaz—muscovite greisens accompanying the quartz lenses of the first stage of mineralization are strongly brecciated. Quartz—cassiterite veins (the second stage of mineralization) form a complex interconnecting network in the greisens. At points of juncture of veinlets there are pocketlike swellings whose thickness is in many cases 10 to 15 times greater than that of the veinlets. Topaz in the brecciated greisens is completely altered to muscovite. The overall strike of the brecciated zone is rigorously east-west. There is a rather thick latticework zone consisting of intersecting veinlets with east-west and northeast-southwest strike in the granite that adjoins the mineralized zone of crushing on the north. The mineral composition of the veinlets of this zone, the alteration of wall-rock minerals to chlorite and hematite, and the truncation of ore bodies of the first and second stages by these veinlets allow us to consider this latticework zone to be a system of veinlets of the third stage of mineralization. Finally, over the entire deposit area there are gangue quartz—fluorite veins of the fourth stage of mineralization; these are restricted to fractures with east-west and northeast-southwest strike formed by renewed opening of the mineralized fractures within the latticework zones of the first stage.

From the foregoing it is seen that the Shakh-Shagaila deposit was formed in the strongly disturbed parts of an ore-bearing intrusion and is characterized by multistage ore deposition. The multistage character of the process was caused by pulsations in tectonic movements; mineralization of the fractures formed in each of the stages differed qualitatively from mineralization of the previous stage. Grigor'ev and Dolomanova (1956) established, for tin-ore deposits of the transitional type, that veinlets of the early stages show mineral associations typical of deposits of the quartz—cassiterite type and veinlets of the later stages show mineral associations typical of deposits of the cassiterite—sulfide type. This pattern is quite evident also for the Shakh-Shagaila deposit, whose deposition was divided into four stages: (1) quartz—cassiterite—topaz, (2) quartz—cassiterite—fluorite, (3) chlorite—sulfide, and (4) quartz—fluorite (gangue). Table 2 gives the mineral associations characteristic of each stage and data on wall-rock alteration.

Table 2

Stage	Morphology of ore bodies	Wall-rock alteration of granite	Mineral associations
Quartz—cassiterite—topaz	Thick lenticular quartz bodies	Quartz, quartz—topaz, topaz—muscovite, and quartz—muscovite greisens	Massive, coarsely crystalline quartz, topaz, cassiterite I, muscovite, fluorite, wolframite, rutile
Quartz—cassiterite—fluorite	System of branching veinlets with variable thickness from 3-4 mm to 10-12 cm	Muscovitization of quartz—topaz greisens, formation of quartz—muscovite greisens (with Fe muscovite)	Comby quartz, coarsely columnar quartz, collomorphic quartz, cassiterite II, Fe muscovite, pyrite, fluorite, adularia, topaz, (chalcopyrite?)
Chlorite—sulfide	Veinlets 0.5 to 2.5 cm thick	Strong replacement of quartz—muscovite greisens by chlorite and hematite; formation of quartz—chlorite and quartz—chlorite—hematite rock	Chlorite, fine-grained quartz, adularia, fluorite, hematite, cassiterite III, calcite, pyrite, chalcopyrite, galenite
Quartz—fluorite	Thick veins of hornfels-like quartz	Quartzification of wall rock	Hornfelslike quartz, comby quartz, fluorite, sericite, hematite

WALL-ROCK ALTERATION

Ore bodies of the various stages of mineralization are accompanied by correspondingly various types of metasomatically altered rock near the veins. Thus, ore bodies of the first stage are accompanied by quartz greisens and quartz—topaz greisens; ore bodies of the second stage are accompanied by quartz—chlorite and quartz—chlorite—hematite rock. Finally, the formation of gangue quartz—fluorite veins of the fourth stage involved slight quartzification of the granite.

Alteration near the veins has been studied in most detail in the region of the main ore zone, where ores of the first three stages of mineralization are most fully developed. As noted above, in the central part of the main ore zone there are lenticular quartz bodies representing the first stage of mineralization, accompanied by a rather thick aureole of altered rock in which certain varieties which progressively replace one another in space can be distinguished.

A section from unaltered wall rock into the vein can be represented as follows: granite, muscovitized granite, quartz–muscovite greisen, quartz–muscovite–topaz greisen, quartz–topaz greisen, quartz–topaz greisen, quartz greisen, vein filling. For ore bodies of the first stage, such a section through the zone around the vein is observed to be ideally developed only in the southern part of the main ore zone. In contrast to the northern part, where superposition of the later stages of mineralization is widespread and the greisens associated with ore bodies of the first stage were altered substantially, superposition is almost completely absent in the southern part, so that it is possible to study the alteration of the granite associated with the first stage of mineralization.

Quartz greisens are immediately adjacent to the vein. The contact of the greisens with the vein quartz is quite sharp, but in places it is somewhat diffuse. The transition is complete over a very narrow zone of 0.5 to 1 cm. Macroscopically the quartz greisens are coarse-grained, light-gray rock with small (2 to 3 mm) insets of fluorite and irregular accumulations of finely crystalline cassiterite. Besides clotlike accumulations, the cassiterite is present also as a uniform dissemination of very small crystals. Also, minor thinly lamellar muscovite, closely associated with fluorite or forming thin latticework segregations in the main quartz mass, is everywhere present in the rocks. Minor disseminated topaz, which is rather evenly distributed in regions of metasomatic quartz as small (0.1 to 0.2 mm), long prismatic crystals, is present in the greisen in addition to muscovite and quartz.

The presence of rare but evenly distributed small cavities with an average size of 3 mm is characteristic of the quartz greisen. The cavities are usually rounded but in some cases irregular. The cavity walls are covered with small quartz crystals; cassiterite crystals closely associated with thinly lamellar muscovite and fluorite are less common. These cavities are almost completely absent in greisen samples taken from pits 5.5 to 6 m deep. Isolated cavities are 80 to 90% filled with fluorite.

Under the microscope the quartz greisen shows clear porphyroblastic texture. The inequigranular nature of the quartz greisen, which is an almost monomineralic quartz rock, is caused by the presence of quartz of two genetic types, relict and secondary. Relict

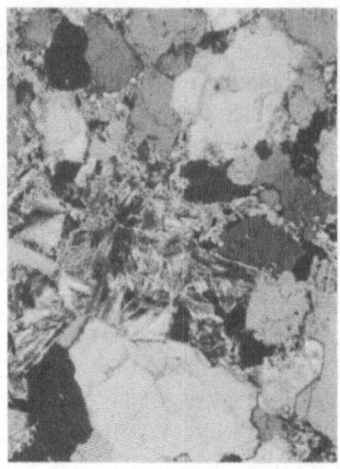

Fig. 70. Quartz greisen with irregular segregations of finely crystalline topaz (in the upper left corner). In the center of the photograph is a cavity filled with muscovite (× 46; crossed nicols).

quartz is present as rare, large grains (3 to 7 mm) that are phenocrysts in the granite and as small, equidimensional grains in the
granite groundmass. These quartz grains show well-defined coarsely
mosaic texture and nonuniform extinction.

Along with complete replacement of feldspars by quartz, the
residual quartz also underwent some alteration. During greisenization,
fringes of secondary quartz formed on the relict quartz grains,
assuming the optical orientation of the original quartz. The original
outlines of most of the relict quartz grains are clearly marked by
discontinuous borders of very fine muscovite lamellae. The quartz
forming the peripheral fringes has uniform extinction and shows no
signs of mechanical deformation (Fig. 70).

In the groundmass, secondary quartz forms a medium-grained
aggregate of irregular grains with uneven and sometimes strongly
curved boundaries. In contrast to the original quartz, in the secondary quartz very small gas and liquid inclusions, which form a complex
network of chains of vacuoles, are very abundant. Minor disseminated acicular topaz crystals and irregular segregations of finely
lamellar muscovite, mostly restricted to the boundaries of quartz
grains, are everywhere present in this quartz. The muscovite segregations are closely associated with rounded fluorite grains that are
evenly distributed in the secondary quartz aggregates. The fluorite
is idiomorphic toward the secondary quartz and forms straight, even
boundaries with the quartz grains. The boundaries between fluorite
and relict quartz are strongly corroded. The relationship between
secondary quartz and cassiterite is contradictory. In some cases

cassiterite is sharply idiomorphic toward the quartz, in other cases cassiterite grains and secondary quartz grains in contact show mutual-growth boundaries, while in still other cases the cassiterite is clearly xenomorphic toward the quartz. Such relations between the cassiterite and secondary quartz are evidence that influx and crystallization of tin dioxide proceeded throughout the entire process of metasomatic alteration of the grains.

Because it is possible to speak of two sources of the silica of the secondary quartz (silica brought in from the outside and silica liberated during alteration of feldspars), the mineral associated in the secondary quartz aggregates is also determined by transported and residual components (cassiterite, fluorite, topaz, muscovite). The thickness of the quartz greisen zone is from 3 to 3.5 m.

Table 3. Chemical Composition of Quartz Greisen

Components	%	Atomic proportions	Atomic proportions of oxygen and fluorine	Number of cations in standard cell
SiO_2	89.18	1486	2972	727
TiO_2	0.05	1	2	0.5
SnO	1.72	11	22	5
Al_2O_3	5.05	100	150	49
Fe_2O_3	0.11	1	1.5	0.5
FeO	—	—	—	—
MnO	0.09	1	1	0.5
MgO	0.22	5	5	2
CaO	1.58	29	29	14
Na_2O	0.12	4	2	2
K_2O	0.32	6	3	3
H_2O^+	0.38	44	44	21.5
H_2O^-	—	—	—	—
F	1.62	—	85 [41.5]*	—
Sum . . .	100.54		3316	825
	0.68		42.5	
	99.86		3273.5	

Formula of rock:

$$K_3Na_2Ca_{14}Mg_2Mn_{0.5}Fe^3_{0.5}Al_{49}Si_{0.5}Sn_5Si_{727}F_{41}(OH)_{21}O_{1538}$$

The following elements were found in the quartz greisen spectrographically in addition to the principal components: Bi (0.00n⁻), Y (0.00n), Ga (0.00n), and Zn (0.00n).

*The number of fluorine ions in the standard cell is shown in brackets.

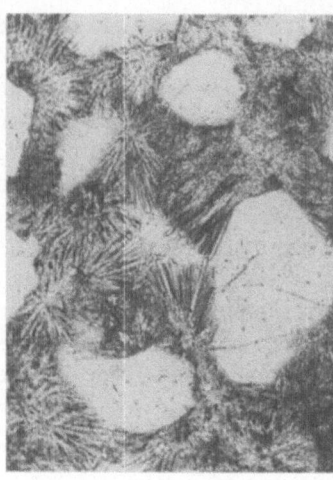

Fig. 71. Quartz—topaz greisen (x 70; uncrossed nicols).

A chemical analysis of the quartz greisen (Table 3) made by N. V. Voronkova has been recalculated by the oxygen method of Barth (1955), as have all subsequently cited analyses of metasomatically altered rocks. Without discussing the method of recalculation (Barth, 1955; Chetverikov, 1956), we note only that certain changes, which are consistent with the principle of the method itself, were made. From the fact that some of the metasomatically altered rocks studied by the writer contain substantial quantities of fluorite, and taking as a basis the closeness of ionic radii of fluorine (1.33 Å) and oxygen (1.36 Å), we assume that the size of the standard cell is determined by the sum of the oxygen and fluorine atoms, made equal to 1600. In recalculation, from the sum atomic proportion of oxygen we subtract the atomic proportion of oxygen corresponding to the correction for fluorine. Making the atomic proportion of fluorine in this analysis 1600, we obtain the number of fluorine atoms in the standard cell. The further recalculation is done in the usual way.

Away from contact with the vein quartz, the amount of topaz in the quartz greisen increases noticeably, and quartz greisen grades into quartz—topaz greisen. Because topaz is the predominant component in this rock, these greisens will be termed topaz greisens. The topaz-greisen zone is nowhere thicker than 1 m.

The topaz greisen is a dense, white rock with well-defined porphyritic aspect caused by the presence of large relict quartz phenocrysts uniformly distributed in the fine-grained topaz. A rather even distribution of finely crystalline cassiterite, forming spotlike segregations in the rock, is characteristic of the topaz greisen.   Dark-violet fluorite and yellowish-green, finely lamellar muscovite are present in the topaz greisen as extremely rare scattered grains. Rare, irregular, small (1.5 by 2 mm) cavities with microreniform topaz aggregates on the walls are present in the greisen.

It was found under the microscope that the fine-grained groundmass of the greisen consists of spherulitic topaz aggregates 0.03-0.04 mm to 0.5-0.6 mm in size filling the spaces between relict quartz phenocrysts (Fig. 71). These are usually irregular in shape because of uneven development of the individual radiating sectors; this in turn is explained by hindered growth of the spherulites. The contact surfaces between spherulites are uneven and sometimes strongly curved.

Cavities filled with secondary quartz are present in the compact spherulitic mass. The dense spherulitic aggregates of topaz that face into these cavities have the form of hemispheres with radiating structure.   Isolated, rigorously spherical spherulites of topaz are observed in the secondary quartz (Fig. 72). The topaz spherulites consist of finely acicular radial crystals which gradually thicken away from the center and which are terminated at the margin of the spherulite by well-formed ends with crystal forms that are indeterminate owing to the small size of the crystals.

In the acicular topaz aggregates forming the spherulites, relicts of finely lamellar muscovite are common. The relict muscovite is

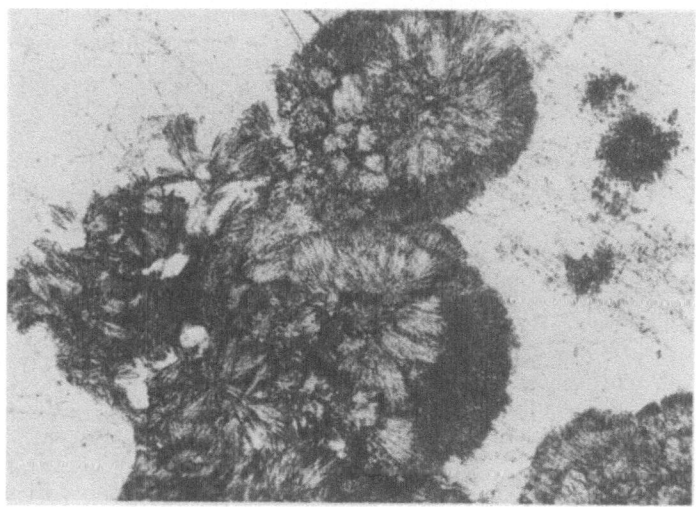

Fig. 72. Topaz spherulites in secondary quartz (x 150; uncrossed nicols).

Table 4. Chemical Composition of Quartz–Topaz
Greisen*

Components	%	Atomic proportions	Atomic proportions of oxygen and fluorine	Number of cations in standard cell
SiO_2	71.56	1192	2384	582
TiO_2	0.04	—	—	—
SnO_2	0.48	3	3	1.5
Al_2O_3	21.72	426	639	208
Fe_2O_3	Trace	—	—	—
FeO	Not determined	—	—	—
MnO	0.01	—	—	—
MgO	0.08	3	3	1.5
CaO	0.43	7	7	3.4
Na_2O	0.07	4	2	2
K_2O	0.51	10	5	5
H_2O^+	0.12	12	12	6
H_2O^-	0.05	—	—	—
F	8.27	—	435 [212]†	—
Sum . . .	103.34		3493	809.4
	3.49		217	
	99.85		3276	

Formula of rock:

$$K_5Na_2Ca_{3.4} \; Mg_{1.5}Al_{208}Sn_{1.5}Si_{582}F_{112}(OH)_6 \, O_{1382}$$

*The following elements were found in the quartz–topaz greisen
spectrographically in addition to the principal components: Bi
(0.0n⁻), Ga (0.00n⁺), Zr (0.00n), Y (traces), and Sr (traces).
†The number of fluorine ions in the standard cell is shown in
brackets.

most abundant in the inner parts of the spherulite; it is absent from
the outer parts.   Boundaries between the finely spherulitic topaz
aggregates and the relict quartz are strongly corroded. The quartz
is in many cases almost completely replaced by finely acicular topaz
aggregates in which only relict, uniformly extinguishing quartz grains
are preserved.   There is much less cassiterite in the topaz greisen
than in the quartz greisen. Fine-grained cassiterite aggregates are
restricted exclusively to the main topaz mass. The boundaries between the cassiterite and the topaz spherulites are in some cases
straight and in others stepped and mutually grown, which is evidence
that these minerals crystallized at the same time.

Table 4 gives a chemical analysis of the quartz–topaz greisen
made by N. V. Voronkova.

The indices of refraction of topaz separated from the rock are $a = 1.608 \pm 0.002$, $\beta = 1.611$, $\gamma = 1.619 \pm 0.002$, and $\gamma - a = 0.11$.

Away from the quartz-greisen zone and toward the granite, topaz greisen grades into topaz–muscovite greisen. In the transitional zone of the topaz greisen, muscovite becomes more abundant and the rock becomes yellowish green. The topaz–muscovite greisen zone is up to 2 m thick.

The topaz–muscovite greisen is a dense, greenish-gray porphyritic rock. The porphyritic aspect of the greisen is masked somewhat by the gray color of the relict quartz phenocrysts. Complete absence of any macroscopically visible cavities is characteristic of the topaz–muscovite greisen, but great numbers of very small (fractions of a millimeter) lenticular segregations and very thin capillary quartz veins associated with fluorite and cassiterite are present throughout the rock.

Under the microscope the rock is seen to be a fine-grained topaz–muscovite aggregate with rare, strongly corroded insets of relict quartz and numerous flaky segregations of finely crystalline cassiterite. Topaz forms irregular accumulations of finely acicular, less commonly spherulitic aggregates that are more or less uniformly distributed in the dense, finely lamellar muscovite mass. Relicts of coarsely lamellar muscovite are present here and there. Topaz strongly corrodes relict quartz, and there is a rather strict correlation between amounts of relict quartz and topaz in the various parts of the rock. Where relict quartz grains are most abundant the groundmass consists mainly of topaz. In the immediate contact zones be-

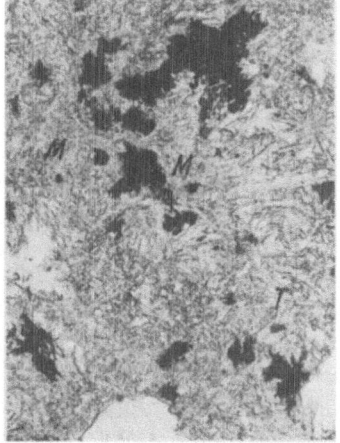

Fig. 73. Topaz–muscovite greisen. Acicular and spherulitic topaz (T) is irregularly distributed in a finely lamellar muscovite mass (M). The black parts are cryptocrystalline cassiterite (× 46; uncrossed nicols).

Table 5. Chemical Composition of Topaz–
Muscovite Greisen

Component	%	Atomic proportions	Atomic proportions of oxygen and fluorine	Number of cations in standard cell
SiO_2	74.45	1239	2478	620
TiO_2	0.03	—	—	—
SnO_2	0.74	5	10	2.5
Al_2O_3	17.71	348	522	174
Fe_2O_3	0.18	2	3	1
FeO	—	—	—	—
MnO	0.05	1	1	0.5
MgO	0.17	3	3	1.5
CaO	1.36	25	25	12.5
Na_2O	0,13	4	2	2
K_2O	2.74	58	29	29
H_2O^+	1.32	144	114	72
H_2O^-	0.06	—	—	—
F	2.79	—	147 [74] *	—
Sum	101.73		3264	911
	1.17		73	
	100.56		3191	

Formula of rock:

$$K_{29}Na_2Ca_{12.5}Mg_{1.5}Al_{174}Sn_{2.5}Si_{620}F_{74}(OH)_{72}O_{1454}$$

The following elements were determined spectrographically in
the topaz–muscovite greisen: Sr (0,n⁻), Bi (0.0n), Ba (0.0n⁻),
Ga (0.00n), Zn (0.00n), Cu (0.00n⁻), and Cr (0.00n⁻).
*The number of fluorine ions in the standard cell is given in
brackets.

tween the topaz aggregates and the corroded relict quartz, the topaz
is more coarsely crystalline (Fig. 73).

Cassiterite, as flaky or, less commonly, granular aggregates, is
quite uniformly distributed in the main topaz–muscovite mass of the
greisen. Cryptocrystalline cassiterite aggregates are found in various
stages of recrystallization. The initial stages are characterized by
the appearance of a flaky aggregate of fine-grained segregations in
the cryptocrystalline mass. Subsequent stages are characterized by
an increase in the size and grain size of these segregations. Commonly
in these intermediate stages of recrystallization several of the
granular segregations grew together to form a granular nucleus in the
central part of the flaky aggregate, occupying about 50 to 60% of the

volume of the aggregate. The final stages of recrystallization of the cryptocrystalline aggregate are represented by coarse-grained crystals, which in some cases still retain a thin fringe of cryptocrystalline cassiterite and in other cases have distinct outlines.

The restriction of the cassiterite aggregates with distinct recrystallization features to substantially topaz parts of the rock is noteworthy. Such features are absent in the flaky segregations of cryptocrystalline cassiterite in the parts of the rock consisting of muscovite. Thus, recrystallization of cryptocrystalline cassiterite is related to topazification of quartz–muscovite greisen during metasomatic growth of the topaz zone.

Table 5 gives the chemical composition of the topaz–muscovite greisen (analysis made by N. V. Voronkova).

Quartz–muscovite greisen that forms the zone between topaz–muscovite greisen and granite is macroscopically almost indistinguishable from the topaz–muscovite greisen. A minor difference is the appearance of yellowish coloring and the more distinct porphyritic character of the quartz–muscovite greisen. The quartz–muscovite greisen zone is up to 1.5 m thick.

Under the microscope the quartz–muscovite greisen is an inequigranular rock whose groundmass consists of a dense aggregate of finely lamellar muscovite containing unevenly distributed phenocrysts of relict quartz and irregular segregations of spherulitic aggregates of coarsely lamellar muscovite (Fig. 74). The thinly lamellar muscovite is developed in the fine-grained granite groundmass, but the segregations of coarsely lamellar spherulitic muscovite are

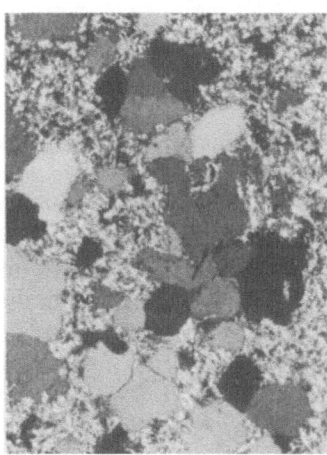

Fig. 74. Quartz–muscovite greisen
(x 46; crossed nicols).

Table 6. Chemical Composition of Quartz–Muscovite Greisen

Component	%	Atomic proportions	Atomic proportions of oxygen and fluorine	Number of cations in standard cell
SiO_2	78.28	1305	2610	635
TiO_2	0.07	1	2	0.5
SnO_2	0.14	1	2	0.5
Al_2O_3	14.50	284	426	138
Fe_2O_3	0.27	4	6	2.5
FeO	0.18	3	3	1.5
MnO	Trace	—	—	—
MgO	0.10	3	3	1.5
CaO	0.20	4	4	2
Na_2O	0.12	4	2	2
K_2O	3.91	82	41	40
H_2O^+	1.72	184	184	90
H_2O^-	0.20	—	—	—
F	0.06	—	5 [2.5] *	—
Sum	99.75		3288	914
$-O = F$	0.03		2	
	99.72		3286	

Formula of rock:

$$K_{40}Na_2Ca_2Mg_{1.5}Fe^2_{1.5}Fe^3_{2.5}Al_{138}Sn_{0.5}Ti_{0.5}Si_{635}F_{2.5}(OH)_{90}O_{1507}$$

The following elements were found spectrographically in the quartz–muscovite greisen: Ba (0.0n), Zr (0.0n⁻), Cr (0.0n⁻), Bi (0.00n), V (0.000n), and Cu (0.000n).

*The number of fluorine ions in the standard cell is given in brackets.

developed mainly in potash feldspar phenocrysts. Relicts of potash feldspar, all with the same orientation, are in some cases present among the spherulitic aggregates of coarsely lamellar muscovite. Relict quartz grains are usually weakly corroded by muscovite.

The muscovite is a pure variety containing no iron; this is confirmed by both chemical analysis (Table 6) and optical data. The average index of refraction of the finely lamellar muscovite is 1.580. The indices of refraction of the coarsely lamellar muscovite are $\alpha = 1.553 \pm 0.002$, $\beta = 1.580$, $\gamma = 1.585 \pm 0.002$; $\gamma - \alpha = 0.032$; and $2V$ measured on a universal stage is 45°.

In the quartz—muscovite greisen there are sparsely disseminated acicular topaz crystals and irregular accumulations of cryptocrystalline cassiterite.

Table 6 gives the chemical composition of the quartz—muscovite greisen (analysis made by O. Ya. Nikolaeva).

Granite in contact with quartz—muscovite greisen is quite strongly replaced by muscovite within a relatively narrow belt (about 1 m). It is seen under the microscope that plagioclase was most strongly replaced and that potash feldspar was only very weakly replaced. It is characteristic that within this belt individual large potash feldspar phenocrysts are replaced by fine-grained aggregates of equidimensional albite grains.

The zone of muscovitized granite gives way to an albitized zone. In this zone the granite loses its strongly porphyritic texture and

Table 7. Chemical Composition of Albitized Granite

Components	%	Atomic proportions	Atomic proportions of oxygen and fluorine	Number of cations in standard cell
SiO_2	78.24	1304	2608	660
TiO_2	0.06	1		0.5
Al_2O_3	12.45	246	369	124
Fe_2O_3	0.06	1	1.5	0.5
FeO	0.40	6	6	3
MnO	0.09	1	1	0.5
MgO	0.04	—	—	—
CaO	0.39	7	7	3.5
Na_2O	4.43	142	71	72
K_2O	3.47	74	37	37
F	None	—	—	—
S	0.06	—	—	—
H_2O^+	0.44	—	—	—
H_2O^-	0.04	56	56	28
Sum	100.17			

Formula of rock:

$$K_{37}Na_{74}Ca_{3.5}Mn_{0.5}Fe_3^2F\,e_{0.5}^3Al_{124}Ti_{0.5}Si_{660}\,(OH)_{28}O_{1572}$$

The following elements were found spectrographically in the albitized granite: Sn (0.0n), Zr (0.0n⁻), Ba (0.0n⁻), Cr (0.00n), Cu (0.00n⁻), Ga (0.00n⁻), and V (0.000n).

Table 8. Rock Porosities

Rock	Specific weight	Volumetric weight	True porosity, %	Effective porosity, %
Granite	2.634	2.599	1.33	1.38
Muscovitized granite	2.636	2.593	1.66	1.69
Muscovite greisen	2.685	2.642	1.60	1.40
Topaz greisen	2.796	2.789	0.25	0.36 *
Quartz greisen	2.721	2.701	0.74	0.61 *

*Repeated determinations of effective porosity gave 0.98 and 1.57% for the topaz greisen and 1.08 and 1.62% for the muscovite greisen.

becomes more even-grained.   Potash feldspar phenocrysts are almost entirely replaced by fine-grained albite.   Potash feldspar in the fine-grained groundmass of the granite is also albitized, though slightly less intensely.   In contrast to the magmatic albite, the secondary albite, which consists of finely tubular, somewhat serpentinized grains, forms aggregates of relatively equidimensional grains that are very fresh.

Table 7 gives the chemical composition of the albitized granite (analysis made by O. Ya. Nikolaeva).

The thickness of the albitized-granite zone varies within wide limits and is everywhere proportional to the total thickness of the greisens accompanying the ore body.   In the cross section being described, the thickness of the albitized zone is about 1.5 m.

Thus, the sharply expressed metasomatic zonation characteristic of infiltrational metasomatism is observed in the alteration aureole in the wall rock for ore bodies of the first stage of mineralization.

To sum up the foregoing we shall analyze the characteristic features of influx and outflux of components in the formation of the greisen zones of the first stage of mineralization.

Because the method of recalculation used in this paper rests upon the constancy of volume of altered and original rocks, the porosity of the granite and the greisens derived from the granite was studied. Table 8 gives data obtained in the Laboratory of Physical-Mechanical Investigations of Rocks of the Institute of Geology of Ore Deposits, Petrography, Mineralogy, and Geochemistry on the porosity of these rocks.

Comparing the data obtained in recalculation of the formulas of the greisens with the formula of the porphyritic granite along which the greisens developed, we obtain the possibility of a quantitative computation of the transport of components during the formation of the various greisens.

1. For the quartz greisen,

$$K_{49}Na_{51}Ca_{2.5}Mg_{1.5}Fe_3^2Fe_2^3Al_{112}Ti_{0.5}Si_{676}(OH)_{17}O_{1583} \ldots \ldots \text{granite}$$

$$K_3Na_2Ca_{14}Mg_2Mn_{0.5}Fe_{0.5}^3Al_{49}Ti_{0.5}Sn_5Si_{727}F_{41}(OH)_{21}O_{1538} \ldots \text{quartz}$$
$$\text{greisen}$$

Influx of ions		Outflux of ions	
Si = 51	OH = 4	Al = 63	Fe^3 = 1.5
F = 41	Mg = 0.5	Na = 49	Fe^2 = 3
Ca = 11.5	Mn = 0.5	K = 46	
Sn = 5			

2. For the quartz–topaz greisen,

$$K_{49}Na_{51}Ca_{2.5}Mg_{1.5}Fe_3^2Fe_2^3Al_{112}Ti_{0.5}Si_{676}(OH)_{17}O_{1583} \ldots \text{granite}$$

$$K_5Na_2Ca_{3.5}Mg_{1.5}Fe_1^3Al_{208}Sn_{1.5}Si_{582}F_{212}(OH)_6O_{1382} \ldots \text{quartz–topaz}$$
$$\text{greisen}$$

Influx of ions		Outflux of ions	
F = 212	Sn = 1.5	Si = 94	OH = 11
Al = 96	Ca = 1	Na = 49	Fe^2 = 3
		K = 44	Fe^3 = 2

3. For the topaz–muscovite greisen,

$$K_{49}Na_{51}Ca_{2.5}Mg_{1.5}Fe_3^2Fe_2^3Al_{112}Ti_{0.5}Si_{676}(OH)_{17}O_{1583} \ldots \text{granite}$$

$$K_{29}Na_2Ca_{12.5}Mg_{1.5}Fe_1^3Al_{174}Sn_{2.5}Si_{620}F_{74}(OH)O_{1454} \ldots \text{topaz–}$$
$$\text{muscovite greisen}$$

Influx of ions		Outflux of ions	
Fe = 74	Ca = 10	Si = 56	Fe^2 = 3
Al = 62	Sn = 2.5	Na = 49	Fe^3 = 1
OH = 55		K = 20	

4. For the quartz–muscovite greisen,

$$K_{49}Na_{51}Ca_{2.5}Mg_{1.5}Fe_3^2Fe_2^3Al_{112}Ti_{0.5}Si_{676}(OH)_{17}O_{1583} \ldots \text{granite}$$

$$K_{40}Na_2Ca_2Mg_{1.5}Fe_{1.5}^2Fe_{2.5}^3Al_{138}Ti_{0.5}Sn_{0.5}Si_{635.5}Fe_{2.4}(OH)_{90}O_{1507} \ldots \ldots$$
$$\ldots \text{quartz–muscovite greisen}$$

Influx of ions		Outflux of ions	
OH = 73	Sn = 0.5	Na = 49	Fe^2 = 1.6
Al = 26	Fe^3 = 0.5	Si = 40.5	Ca = 0.5
F = 2.4		K = 9	

5. For the albitized granite,

$$K_{49}Na_{51}Ca_{2.5}Mg_{1.5}Fe_3^2Fe_2^3Al_{112}Ti_{0.5}Si_{676}(OH)_{17}O_{1583} \cdot \cdot \text{granite}$$

$$K_{37}Na_{74}Ca_{35}Mn_{0.5}Fe_3^2Fe_2^3Al_{124}Ti_{0.5}Si_{660}(OH)_{28}O_{1572} \cdot \cdot \text{albitized granite}$$

Influx of ions		Outflux of ions	
$Na = 23$	$Ca = 1$	$Si = 16$	$Mg = 1.5$
$Al = 12$	$Mn = 0.5$	$K = 12$	$Fe^3 = 1.5$
$OH = 11$			

Thus, on the whole, the greisens accompanying ore bodies of the first stage of mineralization formed with strong outflux of sodium, calcium, and silica and substantial influx of aluminum, fluorine, and water.

The formation of qualitatively different greisens is characterized by the following features:

1. The quartz-greisen zone is characterized by strong outflux of alkali metals and aluminum and influx of silica, fluorine, and tin. Ore-bearing solutions are the source of these latter elements. Highly concentrated ore-bearing solutions enriched in silica by partial desilification of the granite during the concurrent formation of the topaz, topaz–muscovite, and quartz–muscovite greisens are the source of the silica.

2. The topaz-greisen zone is characterized by almost complete outflux of alkali metals, substantial outflux of silica, and influx of aluminum and fluorine. As for the sources of aluminum, it should be borne in mind that the quantity of aluminum ions passing into the metamorphosing solutions during formation of the quartz greisen (second zone) is clearly insufficient to form the topaz greisen. As already noted, 64 aluminum ions per unit volume go into solution in the formation of the quartz greisen, and 96 aluminum ions are taken up in the formation of the topaz greisen. Taking into account that there is also a considerable influx of aluminum in the formation of the next greisen zones, we must necessarily conclude that ore-bearing solutions are another source of aluminum.

3. The topaz–muscovite greisen zone is characterized by almost complete outflux of sodium, somewhat less than in the previous zone, outflux of silica and potash, and considerable influx of aluminum, calcium, fluorine, and water. In mineral composition this zone is transitional from quartz–topaz greisen to quartz–muscovite greisen. The existence of such a transitional zone is probably related to change of the porosity of the granite that was being transformed into greisen. In the granite within this zone there was a band of fine jointing, which

is confirmed by the presence of regularly oriented "blind" veins filled with quartz, fluorite, and cassiterite in the topaz—muscovite greisen.

4. The quartz—muscovite greisen zone is characterized by almost complete outflux of sodium and considerable outflux of silica. Potassium was removed in very small quantities. Water was supplied in great quantities and aluminum in substantial quantities. The hypothesis of Strelkin (1953) that alteration of feldspars to muscovite is caused above all by influx of aluminum and takes place according to the equation

$$KAlSi_3O_8 + 2Al^{3+} + 6(OH)^- = KAlSi_3O_{10}(OH)_2 + 2H_2O$$

is confirmed. From this equation we see that the ratio of inflowing ions is Al : OH = 1 : 3, which corresponds completely with the writer's data, 26(Al) : 73(OH) = 1 : 2.8.

5. The albitized granite zone is characterized by noticeable outflux of silica and potassium and influx of sodium, aluminum, and water. The albitized zone along the greisen margin was formed by the action of deeply metamorphosed greisenizing solutions enriched in alkali metals, particularly sodium, on the granite. The sharp increase in sodium concentration in the greisenizing solutions disrupted equilibrium between the granite and solution and gave rise to a secondary metasomatic process, albitization. Thus, in most cases of intense and complete greisenization, an albitized zone, represented (depending upon the intensity of greisenization) by albitized granite to albitite, should develop along the periphery of the greisenized aureoles.

Granite is completely transformed into quartz—muscovite greisen within the latticework zones of the first stage of mineralization. Topaz—muscovite greisen is less common, and small irregular areas of quartz greisen and quartz—topaz greisen are observed only in isolated cases. In outward appearance, texture, and mineralogy, these greisens differ in no way from the topaz—muscovite and quartz—muscovite greisens, described above, that accompany the thick lenses of the main ore zone.

The transition from greisen to unaltered granite is gradual, in the following pattern: quartz—muscovite greisen, muscovitized granite, albitized granite, unaltered granite.

Table 9 gives the results of a semiquantitative spectrographic analysis of topaz—muscovite and quartz—muscovite greisen from the various latticework zones of the first stage of mineralization. The analyses were made in the spectrographic laboratory of the Kazakhstan Multipurpose Expedition of the Academy of Sciences of the USSR by

Table 9. Results of Spectrographic Analysis of Topaz–Muscovite and Quartz–Muscovite Greisens from the Latticework Zones of the First Stage of Mineralization (in percent)*

Element	Topaz–muscovite greisen						Quartz–muscovite greisen						
	2605	2641	2659	2682	2711	2609	2642	2664	2677	2683	2692	2710	2724
Al	n^+	n^+	n^+	n^+	n^+	n^+	n^+	n^+	n^+	n^+	n^+	n^+	n^+
Si	n^+	n^+	n^+	n^+	n^+	n^+	n^+	n^+	n^+	n^+	n^+	n^+	n^+
Ka	n	n^+	n	n	n	n^+	n^+	n	n^+	n	n^+	n^+	n^+
Na	$0.n^+$	$0.n^-$	$0.n^-$	$0.0n^+$	$0.n^-$	$0.n$	$0.n^-$	$0.n$	$0.0n^+$	$0.n^-$	$0.n^-$	$0.0n^+$	$0.0n^+$
Ca	$0.n^+$	$0.n$	$0.n$	$0.n^+$	$0.n^-$	$0.n^-$	$0.n^-$	$0.n$	$0.0n$	$0.0n^+$	$0.0n$	$0.n^-$	$0.0n^-$
Mg	$0.n$	$0.n^-$	$0.0n^+$	$0.0n$	$0.0n$	$0.n^-$	$0.0n^+$	$0.0n$	$0.n^-$	$0.n^-$	$0.0n^+$	$0.0n^+$	$0.0n^+$
Fe	$0.n^-$	$0.0n$	$0.0n^-$	$0.0n^-$	$0.0n$	$0.n^-$	$0.n$	$0.n^-$	$0.n^+$	$0.n^-$	$0.0n$	$0.0n^+$	$0.0n^+$
Sr	$0.n^-$	$0.n^-$	$0.n^-$	$0.0n$	$0.0n^+$	—	—	$0.00n^+$	—	—	—	—	—
Ba	$0.0n^-$	$0.0n^+$	$0.0n$	$0.0n^-$	$0.0n^-$	$0.0n$	$0.0n^-$	$0.0n^+$	$0.00n$	$0.0n^-$	$0.0n$	$0.00n^+$	$0.0n^-$
Sn	$0.n$	$0.n^-$	$0.0n$	$0.0n^+$	$0.0n^+$	$0.n^-$	$0.n^+$	$0.0n^+$	$0.n^-$	$0.n^-$	$0.0n^-$	$0.n$	$0.0n$
Bi	$0.0n^-$	$0.0n^-$	$0.0n^-$	$0.00n^+$	$0.0n^-$	$0.00n^-$	—	$0.00n$	$0.00n^-$	$0.00n$	$0.00n$	$0.00n^-$	—
Ti	$0.0n$	$0.0n^-$	$0.0n$	$0.0n^-$	$0.0n^-$	$0.0n^-$	$0.0n$	$0.0n$	$0.0n^+$	$0.0n^+$	$0.0n$	$0.0n^-$	$0.0n$
Zr	$0.00n$	$0.00n$	$0.00n$	$0.00n$	$0.00n^-$	$0.0n^-$	$0.0n$	$0.0n$	$0.0n^-$	$0.0n^-$	$0.0n^-$	$0.0n^-$	$0.0n$
Cr	$0.00n^-$	$0.00n^-$	$0.00n$	$0.00n^-$	$0.00n^-$	$0.0n^-$	$0.0n^-$	$0.0n^-$	$0.0n$	$0.0n$	$0.0n^-$	$0.0n^-$	$0.0n$
Ga	$0.00n^-$	$0.00n^-$	$0.00n^-$	$0.00n$	$0.00n^-$	—	—	$0.00n$	—	—	$0.00n^-$	$0.00n^-$	—
Mn	$0.00n$	$0.00n^+$	$0.0n^-$	$0.00n^+$	$0.00n$	$0.00n^-$	$0.00n^+$	$0.00n$	$0.00n^-$	$0.00n^-$	$0.00n^-$	$0.00n^+$	$0.00n$

*n^-, 1 to 3%; n, 3 to 6%; n^+, 6 to 9% or more.

N. P. Sechina. These data show that topaz–muscovite and quartz–
muscovite greisen of the latticework zones and also the corresponding
greisens of the main ore zone are almost completely devoid of iron
(in contrast to greisens of the second stage of mineralization) and
have a variable tin content and a nearly constant association of minor
elements: Sr, Ba, and Bi for topaz–muscovite greisen and Ba, Cr,
and Zr for quartz–muscovite greisen.

Wall-rock alteration associated with ores of the second stage of
mineralization is represented by quartz–muscovite greisen accompanying quartz–cassiterite veins of the brecciated zone. A
considerable part of the brecciated zone is within the northern flank
of the metasomatic aureole of the first stage of mineralization, and
topaz–muscovite greisens of the first stage of mineralization form
the country rock of the veins of the second stage as well as granite.
At contacts with quartz–cassiterite veins of the second stage of
mineralization these greisens were intensely muscovitized. The
quartz–muscovite greisens thus formed differ from those of the first
stage of mineralization by their darker color, the coarse grain size
of the minerals, and the presence of minor but ubiquitous disseminated pyrite.

Under the microscope, quartz–muscovite greisen of the second
stage shows porphyritic texture, but in contrast to similar greisen of

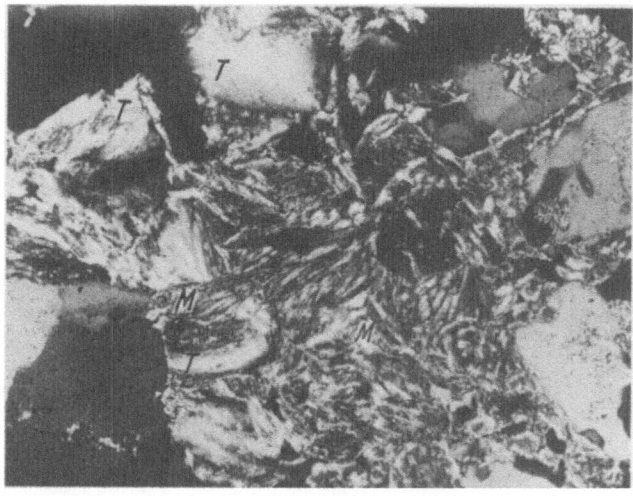

Fig. 75. Quartz–muscovite greisen of the second stage with
residual topaz (T), partially replaced by muscovite (M) (× 70;
crossed nicols).

Table 10.   Chemical Composition of Quartz–
Muscovite Greisen of the Second Stage of
Mineralization

Component	%	Atomic proportions	Atomic proportions of oxygen	Number of cations in standard cell
SiO_2	79.50	1325	2650	645
TiO_2	0.08	1	2	0.5
Al_2O_3	13.20	258	387	126
Fe_2O_3	0.00	—	—	—
FeO	1.13	15	15	7
MnO	0.11	1	1	0.5
MgO	0.13	3	3	1.5
CaO	0.05	2	2	1
Na_2O	0.12	4	2	2
K_2O	3.48	74	37	36
H_2O^+	1.74	184	184	90
H_2O^-	0.00	—	—	—
F	0.00	—	—	—
S	0.10	—	—	—
Sum	99.64		3283	909.5

Formula of rock:

$$K_{36}Na_2Ca_1Mg_{1.5}Mn_{0.5}Fe_7^{2}Al_{126}Ti_{0.5}Si_{645}O_{1510}(OH)_{90}$$

The following elements were identified in the quartz–muscovite greisen spectrographically: Sn $(0.0n^+)$, Zn $(0.0n^-)$, Ba (0.00n'), Y (0.0n'), La (0.01), Cr (0.00n'), Cu (0.00n), Ga (0.00n'), Be (0.001), Mo (0.000), and V (0.000).

the first stage the groundmass, consisting of coarsely lamellar mus-covite, is more even grained.

Relicts of spherulitic topaz aggregates are in some places observed in the quartz–muscovite greisen (Fig. 75). The darker color of the groundmass muscovite is probably caused by an admixture of the ferrophengite molecule $K_2Fe^{2+}Al_3(OH)_4Si_7AlO_{20}$. The indices of refraction of the muscovite are $\alpha = 1.564 \pm 0.002$, $\beta = 1.592$, and $\gamma = 1.598 \pm 0.002$.

Pyrite, as small (up to 1 mm) cubic crystals, is evenly dissemi-nated in the rock. Crystal forms other than {100} are not observed. The cube faces are covered by fine striations. Spectrographic analysis of pyrite separated from the rock reveals the following elements besides the main components: Si $(0.0n^+)$, Al $(0.0n^-)$, Na $(0.0n^-)$, Mn $(0.0n^-)$, Cu $(0.00n^+)$, Pb (0.00n), Sn $(0.00n^-)$, V (0.000n), and Bi (traces of lines).

Table 10 gives the chemical composition of quartz–muscovite greisen of the second stage of mineralization formed by muscovitization of topaz–muscovite greisen of the first stage (analysis made by O. Ya. Nikolaeva).

The quartz–muscovite greisen accompanying the quartz–cassiterite veins of the second stage but present in the granite differs from those described above by the absence of relict topaz and by the presence of more densely dessiminated pyrite. Table 11 gives the chemical composition of this greisen (analysis made by O. Ya. Nikolaeva).

Table 11. Chemical Composition of Quartz–Muscovite Greisen of the Second Stage of Mineralization (along Granite)

Component	%	Atomic proportions	Atomic proportions of oxygen	Number of cations in standard cell
SiO_2	76.33	1270	2540	625
TiO_2	0.06	1	2	0.5
SnO_2	0.12	1	2	0.5
Al_2O_3	14.06	276	414	136
Fe_2O_3	0.00	—	—	—
FeO	2.46	35	35	17
MnO	0.24	3	3	1.5
MgO	0.19	5	5	2.5
CaO	0.29	5	5	2.5
Na_2O	0.12	6	3	3
K_2O	3.47	74	37	36
H_2O^+	0.79⎱	166	166	81
H_2O^-	0.66⎰			
F	0.00	—	—	
S	0.83	26	36 *	13†
Sum	99.62		3248	918.5
$0 = S$	—0.41		—25	
	99.21		3223	

Formula of rock:

$$K_{36}Na_3Ca_{2.5}Mg_{2.5}Mn_{1.5}Fe^2_{17}Al_{136}Ti_{0.5}Si_{625}O_{1506}S_{13}(OH)_{81}$$

The following elements were found spectrographically in muscovite greisen of the second stage developed along granite: Zr (0.0n⁻), Ba (0.0n⁻), Y (0.0n⁻), Cu (0.0n⁻), Cr (0.0n⁻), Mo (0.00n⁻), Ga (0.00n⁻), Be (0.00n⁻), and V (0.000n).

*Number of oxygen atoms equivalent in volume to 26 sulfur atoms.
†Number of sulfur atoms in the standard volume.

In comparing the formulas of the muscovite greisen of the second stage of mineralization with the formulas of the rocks in which they formed, we obtain the following data on the quantitative transport of components:

1. $K_{49}Na_{51}Ca_{2.5}Mg_{1.5}Fe_3^2Fe_2^3Al_{112}Ti_{0.5}Si_{676}O_{1583}(OH)_{17}$ . . . . . .granite

$K_{36}Na_3Ca_{2.5}Mg_{2.5}Mn_{1.5}Fe_{17}^2Al_{136}Ti_{0.5}Sn_{0.5}Si_{625}O_{1506}(OH)_{81}$ . .muscovite greisen

Influx of ions		Outflux of ions	
OH $= 64$	Mn $= 1.5$	Si $= 51$	K $= 13$
Al $= 24$	Mg $= 1$	Na $= 48$	Fe$^3 = 2$ *
Fe$^2 = 14$ *	Sn $= 0.5$		
S $= 13$			

2. $K_{29}Na_2Ca_{12.5}Mg_{1.5}Fe_1^3Al_{174}Sn_{2.5}Si_{620}O_{1441}(OH)_{72}F_{87}$ . . topaz– muscovite greisen

$K_{36}Na_2Ca_1Mg_{1.5}Mn_{0.5}Fe_7^2Al_{126}Ti_{0.5}Si_{645}O_{1510}(OH)_{90}$ . . muscovite greisen

Influx of ions		Outflux of ions	
Si $= 25$	Fe$^2 = 7$	F $= 87$	Sn $= 2.5$
OH $= 18$	Mn $= 0.5$	Al $= 48$	Fe$^3 = 1$
K $= 7$		Ca $= 11.5$	

These data show that the formation of muscovite greisen of the second stage of mineralization is characterized by strong influx of water and aluminum and substantial influx of iron and sulfur. It should be borne in mind that the basic pattern of quantitative redistribution of the main components during formation of the muscovite greisen remains valid irrespective of the stage of mineralization.

From the data given above, it follows that during muscovitization of the acidic granite (irrespective of the deposit and the stage of mineralization) the numbers of ions of the principal components entering and leaving a unit volume are of the following orders:

Influx of ions	Outflux of ions
OH — 70	Na — 50
Al — 25	Si — 45
	K — 10

Muscovitization of the topaz greisen and topaz–muscovite greisen of the first stage took place with strong outflux of aluminum and fluorine and substantial influx of water and silica.

*In addition to influx of Fe^2, there is also reduction of Fe^3 to Fe^2.

Wall-rock alteration associated with ores of the third stage of mineralization (latticework zones in the northern part of the ore area) is expressed as intense chloritization of the granite and, in part, of the muscovite greisen of the second stage of mineralization.

Within the lattice work zones of the third stage of mineralization, granite is completely transformed into quartz–chlorite rock. This rock is represented by dense, fine-grained, dark-green, and greenish-black varieties. The color of the dark varieties is caused by the presence of hematite. In some places the chlorite rock has a distinctive variolitic appearance caused by numerous small (2 to 5 mm) rose-colored adularia crystals.

Under the microscope the light varieties of chloritic rock show inequigranular texture. The main mass of the rock consists of finely lamellar chlorite in which irregular relict quartz crystals are unevenly distributed. Generally, the relict quartz is strongly corroded by the chloritic groundmass (Fig. 76).

The chlorite is a dark-green variety with high index of refraction, 1.665. The dark color and the high index of refraction show that it is an iron-rich chlorite of daphnite type. This is confirmed by the chemical analysis of the chlorite–hematite rock (Table 12).

The most widespread dark varieties of chlorite rock differ from those noted above by the presence of abundantly disseminated hematite,

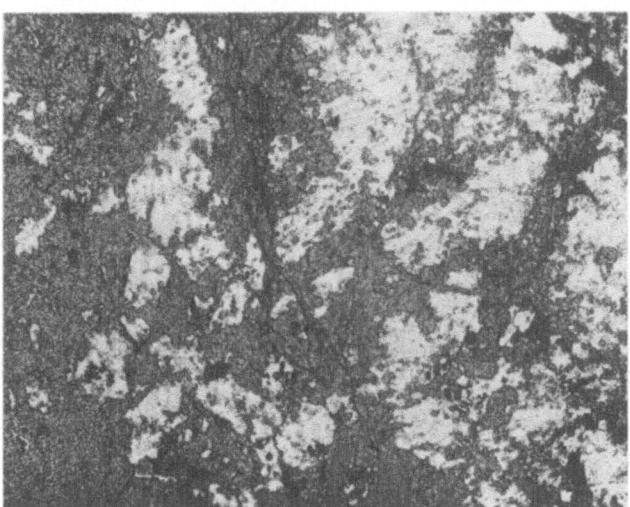

Fig. 76. Quartz–chlorite rock (third stage of mineralization). Residual quartz is strongly corroded by finely lamellar chlorite (x 46; uncrossed nicols).

Table 12.　Chemical Composition of Chlorite–
Hematite Rock

Component	%	Atomic proportions	Atomic proportions of oxygen	Number of cations in standard cell
SiO_2	27.65	461	922	264
TiO_2	0.17	3	6	2
SnO_2	0.11	1	2	0.5
Al_2O_3	11.83	232	348	133
Fe_2O_3	29.31	367	550	210
FeO	21.16	294	294	168
MnO	0.88	13	13	7
MgO	2.81	69	69	39
CaO	0.10	2	2	1
Na_2O	0.12	4	2	2
K_2O	0.91	20	10	11
H_2O^+	5.18	578	578	331
H_2O^-	0.52	—	—	—
F	0.00	—	—	—
S	0.02	—	—	—
Sum	100.77			1168.5

Formula of rock:

$$K_{11}Na_2Ca_1Mg_{39}Mn_7Fe^2_{168}Fe^3_{210}Al_{133}Ti_2Sn_{0.5}Si_{264}O_{1264}(OH)_{331}$$

The following elements were found in this rock spectrographically in addition to the principal components: Zn $(0.n^-)$, Cu and Y $(0.0n^-)$, V $(0.00n^-)$, Be $(0.000n^-)$, and Pb (traces of lines).

which is more or less evenly distributed in the groundmass as fine lamellar crystals. Hematite spherulites and pseudomorphs of hematite after pyrite are common. The pseudomorphs are most often found in chlorite–hematite rock developed along quartz–muscovite greisen of the second stage of mineralization. In addition to iron, the following elements were found spectrographically in hematite separated from the rock: Si, Al, Mg $(0.n^+)$; Mn $(0.0n)$; Ca, V, Sn $(0.0n^-)$; Ti $(0.00n^+)$; Y, Ga (traces of lines).

Rose-colored adularia crystals represented by usually irregular combinations of several wedge-shaped crystals (Fig. 77) are everywhere present in the fine-grained chloritic mass of the chlorite–hematite rock; isolated crystals of adularia are less common. The size of the adularia varies from fractions of a millimeter to 3-4 mm. There is a distinct pattern to their distribution. They are restricted exclusively to the outer zones along the chloritization front of the granite. Adularia in the chlorite–hematite rock becomes noticeably less abundant away from the contact with the granite, and the adularia crystals are more and more corroded and are finally almost entirely replaced by chlorite (Fig. 78).

Table 12 gives the chemical composition of the chlorite–hematite rock (analysis made by O. Ya. Nikolaeva). Comparison of the formula of the chlorite–hematite rock with the formula of the granite shows the following pattern of redistribution of the main components:

$$K_{49}Na_{51}Ca_{2.5}Mg_{1.5}Fe_3^2Fe_2^3Al_{112}Ti_{0.5}Si_{676}O_{1583}(OH)_{17} \qquad \text{granite}$$

$$K_{11}Na_2Ca_1Mg_{39}Mn_7Fe_{210}^2Al_{133}Ti_2Sn_{0.5}Si_{264}O_{1269}(OH)_{331} \qquad \text{chlorite–hematite rock}$$

Thus, formation of chlorite–hematite rock of the third stage of mineralization is characterized by strong influx of water, iron, and magnesium and considerable influx of aluminum and manganese; very great amounts of silica and alkali metals were removed.

It should be noted that wall-rock alteration in the third stage of mineralization in the Shakh-Shagaila deposit repeats the main features of wall-rock alteration characteristic of tin deposits of the chlorite–sulfide type, for which, according to the data of Radkevich (1952), a sharp predominance of iron over magnesium in the chlorites forming the quartz–chlorite rock and the universal presence of such elements

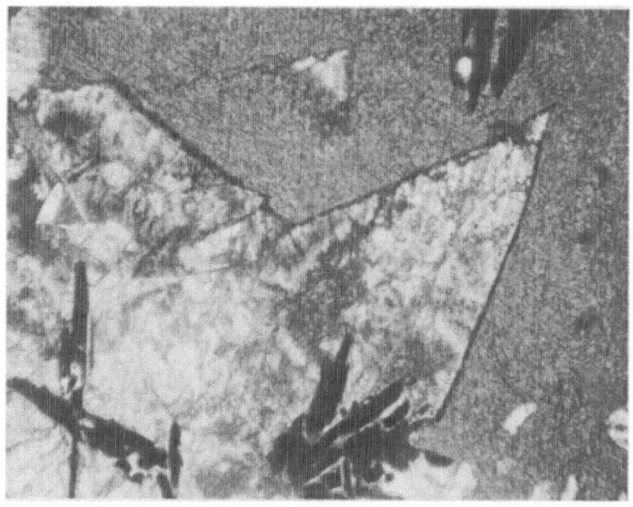

Fig. 77. Gray: intergrowths of wedge-shaped adularia crystals in a fine-grained chlorite mass. Black lamellar crystals: hematite (x 46; uncrossed nicols).

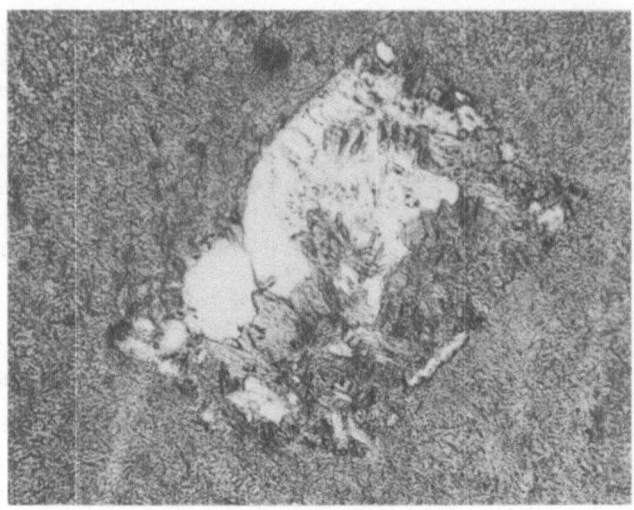

Fig. 78. Structure of replacement of adularia by chlorite (x 46; uncrossed nicols).

as zinc, copper, lead, and other chalcophile elements in these rocks are typical.

Wall-rock alteration of the granite associated with the vein series of the fourth type of mineralization is expressed as minor alteration to serpentine and quartz. The gangue quartz—fluorite veins are every-where accompanied by thin zones of quartzified and serpentinized granite. The thickness of the metasomatic zones varies from 2-3 to 8-10 cm. There is a zone of metasomatic quartz immediately ad-jacent to the vein. This is light gray, sometimes light brown, and has fine-grained texture. It contains a great number of relict potash-feldspar crystals and dustlike hematite lamellae. Away from the vein the metasomatic quartz grades into a grayish-yellow quartz—serpentine rock, which in turn grades into serpentinized granite. This quartzification of the wall rock bears the character of low-temperature hornfels development characteristic of many fluorite deposits of meso- to epithermal type.

Thus, data on wall-rock alteration of the granite for the various types of mineralization show that the nature of the metasomatic altera-tion of the wall rock during formation of the deposits differs sharply in time, reflecting differences in genetic types of ores caused by different stages of mineralization. Quartz—topaz and topaz—muscovite greisens of the first stage of mineralization are characteristic of tin deposits of cassiterite—quartz type. In the Shakh-Shagaila deposit these greisens accompanied the earliest, high-temperature ores. Quartz–chlorite and chlorite–hematite rock of the third stage of mineralization is the characteristic type of wall-rock alteration for tin-ore deposits of the cassiterite–sulfide type. In the Shakh-Shagaila deposit this rock is developed within latticework zones of the late, lowest-temperature ore stage. Muscovite greisen of the second stage of mineralization is characteristic of neither the first nor the second type of tin-ore deposit and can be considered a transitional type of contact metasomatic rock characteristic of intermediate stages of mineralization.

MINERALOGY

Though the mineralogy of the Shakh-Shagaila deposit is distinguished neither by great variety of mineral species nor by complexity of mineral associations, the deposit is of great mineralogical interest because part of its ore shows direct features of colloidal origin. The following features make it possible to bring to light certain patterns for the entire range of the still insufficiently studied processes: (1) the great diversity in morphology and texture of the mineral aggregates present in this deposit; (2) the presence of almost all textural varieties of spherical aggregates; and (3) the great number of textural-morphological types of ores transitional from ores with metacolloidal textures and structures to ores with crystalline-granular and drusy textures and structures. Among these processes are collective crystallization in gels, formation of the various textural types of spherical aggregates, and recrystallization of metacolloidal aggregates. Ores of the second type of mineralization are of special interest in this respect.

It was noted above that the Shakh-Shagaila deposit and all other known tin deposits of transitional type are characterized by multistage ore deposition (see Table 2). Because superposition of later ores on earlier ores is weakly developed in the deposit and thus the specific mineralogical features of each of the stages of mineralization are preserved, the mineralogy of the deposit is described separately for each stage.

Mineralogy of Ores of the First Stage of Mineralization

There are two morphological types of ores of the first stage of mineralization within the deposit area: (1) veins of the thick lattice-

Fig. 79. Relationships of the principal minerals in the vein. Black and dark gray, cassiterite; light gray in center, fluorite (F); white, quartz. Wall rock, topaz—muscovite greisen (x 46; uncrossed nicols).

work zones cutting the ore area in east-west and northwest-southeast direction; (2) lenses in the axial part of the zone of fracturing in the central part of the ore area. The mineral composition of the vein filling is the same for both types:

Predominant	Quartz
Very abundant	—
Moderately abundant	Cassiterite, fluorite
Minor	Topaz, muscovite
Rare	Rutile, wolframite

The granite wall rock was altered to the following varieties of greisen:

1. Quartz greisen
2. Quartz—topaz greisen
3. Topaz—muscovite greisen
4. Quartz—muscovite greisen

Quartz—topaz and topaz—muscovite greisen are predominant within the latticework zones, and quartz greisen and quartz—topaz greisen are almost absent, being found only in isolated cases as small areas. On the other hand, the greisens accompanying the thick lenses include all the varieties listed above, and, because the region of the main ore zone is the most highly disturbed (and thus also the region of most intense activity of metamorphosing solutions on the surrounding granite), there is distinct metasomatic zonation in the greisen aureole.

Veins of the Latticework Zones

The vein type of ores of the first stage of mineralization are widespread in the deposit. They are represented by series of en échelon latticework zones formed by two systems of veins, with east-west and northwest-southeast strike. There are ten latticework zones, with great thickness (from 2 to 60 m) and considerable extent along strike (up to 140 m), and also about twenty-five thin (approximately 0.2 m) zones which, in plan, form a series of lenticular chains, located mainly in the southern periphery of the ore area.

The sameness of mineral composition of the quartz veins is noteworthy. The great majority of the veins forming the latticework zones contain neither cassiterite nor any other ore mineral, and only a small part of them, not more than 10% of the total number, contain minor cassiterite and very rare rutile; of the gangue minerals, fluorite, muscovite, and topaz are present (Fig. 79).

The veins are mostly 1 to 5 mm thick. Isolated veins up to 2 mm thick are considerably less common. Vein fahlbands are generally indistinct. Blending of the vein quartz with the surrounding quartz—muscovite (less commonly topaz—muscovite) greisens makes the vein boundaries indistinct.

Quartz. The quartz that fills the veins is a dark-gray, compact variety. Under the microscope it is seen to be equigranular. The grain outlines are irregular. In places the quartz grains are partially crushed and cut by numerous cracks as a result of cataclasis. In these areas the extinction of the grains is nonuniform and cloudy. Gas and liquid inclusions forming complex networks of chains are present in the quartz grains.

In veins enriched in cassiterite and fluorite, the quartz grains have more distinct, straight-lined outlines. The relations between quartz and cassiterite are contradictory; in most cases the quartz—cassiterite boundaries show a distinctive stepwise character, which corresponds to the presence of mutually grown boundaries. Less commonly the quartz is xenomorphic toward the cassiterite. Finally, in isolated cases the reverse relationship holds. The following minor elements were identified spectrographically in quartz that contained no inclusions visible through a binocular microscope: Fe $(0.0n)$; Ti, Al, Mg $(0.0n^-)$; Sn, W, Mn $(0.00n)$; Cu $(0.00n)$; and Mo $(0.000n)$.

Fluorite. Fluorite is considerably less common than quartz and is usually closely associated with cassiterite. Under the microscope in transmitted light it is light violet. It contains a great number of unordered gas and liquid inclusions. It is sharply xenomorphic toward quartz and cassiterite. The following minor elements were identified spectographically in the fluorite: Si, Al (0.0n); Fe, Y (0.00n$^+$); Mg (0.00n); and Sn, Mn (0.00n$^-$).

Muscovite. Muscovite is found as very small, isolated lamellae unevenly distributed in the quartz. Less commonly it forms dense, finely lamellar aggregates closely associated with fluorite.

Topaz. Topaz is occasionally found as isolated acicular crystals that are sharply idiomorphic toward quartz. It is not found associated with cassiterite.

Cassiterite. Cassiterite is unevenly distributed in the veins. In some cases it almost entirely fills the veins, but in other cases it is disseminated in the quartz as very small crystals. The cassiterite crystals have short prismatic habit (Fig. 79). The following crystal forms were noted for the few crystals that were studied: $m\{100\}$ and $s\{111\}$ and, less common and weakly developed, $a\{010\}$ and $z\{231\}$. The maximum size of the crystals is 3 to 4 mm along the principal axis and 2 to 3 mm across. Crystals not longer than 1 m along the principal axis are most common. The faces of the prisms are uneven and nodular. The faces of the pyramids are mirrorlike, and growth triangles are hardly noticeable. The cassiterite is brownish black to black. Under the microscope in transmitted light it is brownish yellow to reddish brown. Zoning of colors is characteristic. The centers of the crystals are commonly homogeneous and brownish yellow, and only in the outer parts are several dark-brown or reddish-brown zones observed. The color zoning is probably caused by inclusions of very small crystals of wolframite, rutile, and minor columbite, distributed along the zones of growth of the cassiterite crystals; these are characteristic mineral admixtures in cassiterite from deposits of the cassiterite—quartz type (Grigor'ev and Dolomanova, 1951).

The relations between cassiterite and quartz were noted above. The presence of mutually grown faces is evidence of concurrent growth of quartz and cassiterite. Cassiterite is sharply idiomorphic toward fluorite.

Lines of the following elements were noted in the cassiterite spectrographically: Sn (very strong); Fe, W (strong); Ti, Si, Al, Ca (moderate); Sc, Mn (weak); Mg, Bi (very weak); and Ga, Nb, Sb, and Be (traces of lines). Thus, the analysis confirms the possibility of the presence of typical mineral admixtures in this cassiterite: wolframite, rutile, and (in negligible quantities) columbite, characteristic of cassiterites from deposits of the cassiterite–quartz type.

R u t i l e . Rutile is found in the quartz as very thin needles with bright interference colors. They are very unevenly distributed and rare.

Lenses of the Main Ore Zone

Thick quartz bodies of the main ore zone are another less wide-spread type of ore of the first stage of mineralization. In the axial part of the zone of crushing there are five lenticular quartz veins of considerable thickness arranged in a chain. The strike of the vein series is roughly east–west; the azimuths range from 278 to 286° NW. The veins dip to the north at angles of 76 to 84°. The extent of each of the veins is as follows:

Vein no.	Extent, m	Maximum thickness, m
1	$18\frac{1}{2}$	2.0
2	15	3.5
3	10	1.0
4	35	2.6
5	24	2.0

Veins 2, 4, and 5 are interconnected by thin veinlets, but veins 1 and 3 are isolated from the others. As noted above, the veins are characterized by lenticular shape in plan. Some of them have more complex outlines, caused by the presence of thick apophyses, as, for example, vein 4, for which two apophyses (northern and southern) of considerable extent and great thickness are noted. The veins wedge out more or less smoothly. In places of wedging out there are a great number of thin veinlets branching off from the main vein. These veinlets grade into latticework zones in the western and eastern parts of the zone of crushing. From field observations the vein contacts are in some places sharp but in other places diffuse, with the vein quartz seemingly merging into quartz greisen within a very narrow contact zone (approximately 1 cm). Study under the micro-scope shows that the contact between vein quartz and quartz greisen is very sharp (Fig. 80), and the impression of gradual transition in field observations is explained by the presence of a great number of very small slivers of finely crushed and quartzified granite in the vein quartz. The texture of the veins is massive.

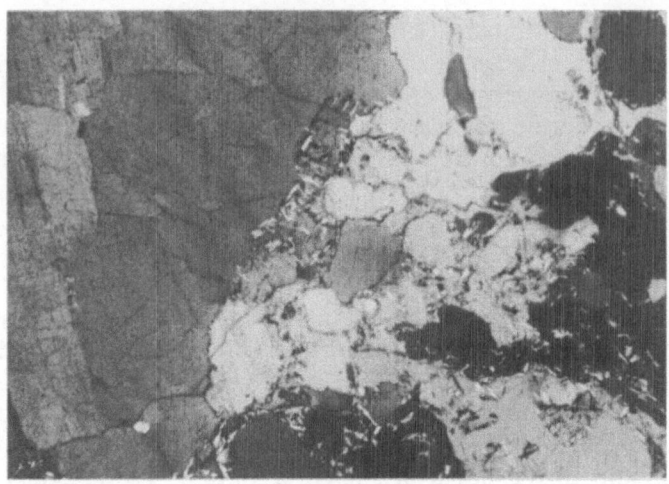

Fig. 80. Nature of the contact between vein quartz and quartz greisen
(x 46; crossed nicols).

The granite near the veins is very strongly altered to quartz, quartz–topaz, quartz–muscovite, topaz–muscovite, and quartz–muscovite greisens.

The mineral composition of the veins is similar to that of the veinlets of the latticework zones. Minor differences are the more nearly universal dissemination of topaz and muscovite and the higher cassiterite content in the lenticular quartz veins.

Quartz. The quartz forming the veins is coarse grained, compact, and grayish white. Under the microscope it shows more or less equigranular texture. The grain boundaries are sharp, but the nature of the boundaries varies from place to place in the vein. Near vein contacts the grain boundaries are straight, and the grain outlines are elongate, sometimes distinctly euhedral. Away from the contacts the grain boundaries become uneven and denticulated, and the grain outlines become grounded or irregular.

The quartz contains great numbers of gas and liquid inclusions, which are unordered in grains near the vein contacts but form radial chains in the outer parts of grains in the inner part of the vein. Concentric chains of inclusions are, in places, also present in these latter grains. The extinction of the grains is very nonuniform. At incomplete extinction the quartz grains in the vein contact zones display a block structure which is characterized by comby texture (Fig. 81). The quartz grains forming the inner parts of the vein show more complex extinction. The centers of the grains have sharp and uniform extinction, and, upon incomplete rotation of the stage, in the outer parts of the grains there are regions that extinguish both in phase and out of phase with the central core. Thus, relicts of radiating structure are clearly evident in the outer zones of the grains. This is evidence of intense recrystallization of the vein quartz and sharp change in the appearance of the original aggregates in the quartz, which are preserved only as relicts.

Below, in describing the minerals of the second stage of minerali-zation, we shall analyze some of the patterns of recrystallization of the vein quartz.

The following minor elements were identified spectrographically in the quartz: Al (0.0n); Fe, Ti, Sn, Mg (0.0n⁻); W, Mn, Na (0.00n); Mo, Bi (0.000n); Cu (traces of lines). Comparing these data with the spectrographic analysis of quartz from the veinlets of the latticework zones, we see that for the vein quartz of the first stage of mineraliza-tion such minor elements as Ti, Sn, Fe, Mn, and W, caused by a mechanical admixture of finely crystalline (and perhaps also finely dispersed) rutile, cassiterite, and wolframite, are most characteristic.

Fluorite. Fluorite is present as irregular, dark-violet insets unevenly distributed in the vein quartz. The fluorite segregations are most abundant in the inner parts of the veins; near the contacts they are rare.

Under the microscope the fluorite insets are seen to be aggre-gates of more or less equidimensional grains 0.5 to 1 mm in size; grains 3 to 4 mm in size are less common. Grain boundaries are in

Fig. 81. Comby-mosaic internal structure of quartz grains from near the vein contacts (x 46; crossed nicols).

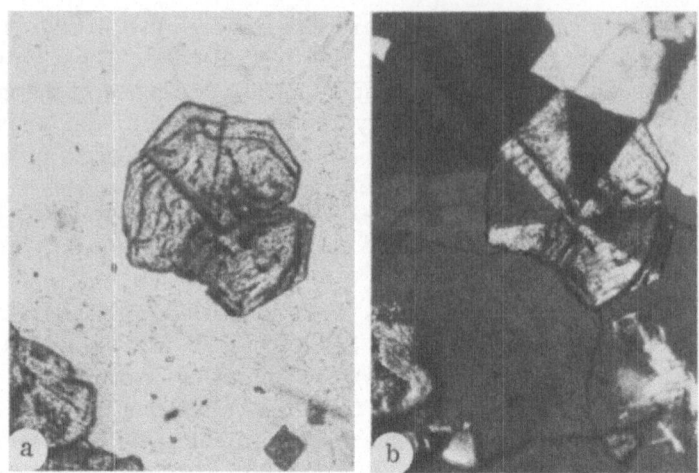

Fig. 82. The most widespread type of cassiterite twins in lenticular quartz bodies of the first stage of mineralization (× 70). (a) Uncrossed nicols; (b) crossed nicols.

most cases stepped, corresponding to mutually grown faces. In transmitted light the fluorite shows well-defined mottled coloring: dark-violet clots are chaotically distributed on a colorless or light-violet background. Color zoning is less common; in this, the restriction of the violet coloring to series of very fine tubular gas and liquid inclusions arranged parallel to the boundaries of the octahedron, with each tubular inclusion oriented perpendicular to these boundaries, is clearly evident. In addition to the very fine tubular inclusions, there are larger gas and liquid inclusions with irregular rounded or amoeboid shape of unordered distribution. By visual observation, the gas phase occupied 25 to 30% and the liquid phase, 70 to 75% of the total volume of inclusions.

Boundaries between fluorite and quartz are in most cases uneven and corroded. Near the vein contacts, fluorite is segregated into interstices between quartz grains, so that its boundaries with the quartz are sharper and straighter, but always with signs of corrosion of quartz by fluorite. Fluorite is xenomorphic toward cassiterite, but in some cases mutually grown boundaries are observed. Fluorite forms intergrowths with muscovite. It is not found in close association with topaz.

Besides the main components, the following components were identified in the fluorite spectrographically: Si, Al $(0.0n^{+})$; Na $(0.0n)$; Fe, Mg $(0.0n^{-})$; Sn, Y $(0.00n^{+})$; and Mn $(0.00n^{-})$. Thus, a substantial iron and magnesium content, a small yttrium content, and an entirely negligible manganese content is characteristic of fluorite of the first stage of mineralization. The markedly elevated sodium content is characteristic of fluorite containing a great number of gas and liquid inclusions.

Muscovite. Muscovite as coarsely lamellar aggregates forms irregular accumulations at the boundaries of quartz grains. It is considerably less common than fluorite. Boundaries between muscovite and quartz are uneven and wavy; muscovite is clearly xenomorphic toward quartz. Muscovite in some cases forms fringes around isolated, well-formed cassiterite crystals; the boundaries of the cassiterite are distinct and even. Relations between muscovite and fluorite are somewhat contradictory; in some cases fluorite is segregated into the interstices of fan-shaped muscovite aggregates, but in other cases coarsely lamellar muscovite aggregates fill the spaces between fluorite grains having clearly euhedral outlines.·

Lines of the following elements were identified in the muscovite spectrographically: K, Al, Si (very strong): Na, Ca (moderate); Mg, Ba (weak); Fe, Mn, Sn (very weak); and Li (traces of lines).

Topaz. Topaz, as individual well-formed crystals with long prismatic (acicular) habit, is considerably less common than muscovite in the vein quartz. Spherulites and paniculate and irregular finely acicular aggregates of topaz are found in places.

Topaz is unevenly distributed in the vein; it is most abundant near the vein contacts but is scarce in the central parts of the vein (and in some veins it is absent altogether). Topaz is sharply idio-

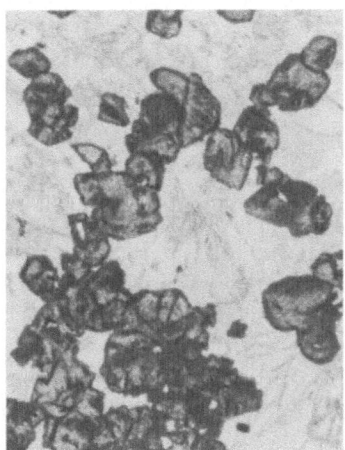

Fig. 83. Idiomorphic segregations of cassiterite in quartz, the central part of the vein filling (× 46; uncrossed nicols).

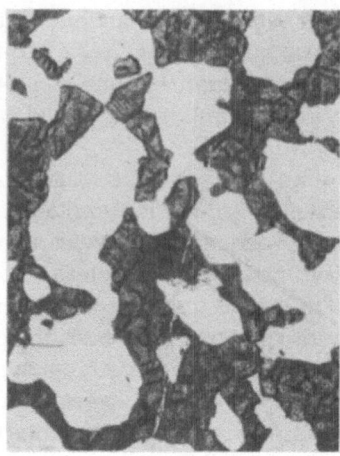

Fig. 84. Xenomorphic segregations of cassiterite in quartz, near the vein contacts (x 46; uncrossed nicols).

morphic toward quartz and cassiterite. Replacement of topaz spherulites by muscovite is clearly apparent.

Cassiterite.   Cassiterite, forming a very fine and even dissemination in the vein quartz and also present as clotlike segregations distributed in clusters in the quartz, is moderately abundant. The cassiterite that is evenly disseminated in the quartz is in the form of small (0.1 to 0.5 mm) well-formed crystals. These crystals are mainly short prismatic, less commonly dipyramidal. As with cassiterite from the veinlets, the forms $m\{110\}$ and $s\{111\}$ are characteristic. A large number of twins, of which geniculated twins on $\{011\}$ are best developed, are observed under the microscope. More complex forms (fourlings and sixlings, Fig. 82) are often observed also; sixling twins were identified by Stulov (1953) in cassiterite from alluvial placers of the northeastern USSR. Stulov's study of complex interpenetration twins showed that macroscopically observable cassiterite fivelings are in most cases actually interpenetration twins of six individuals, and this is confirmed by optical study of these intergrowths.   The sixlings observed by the present writer also give the impression of fivelings because of the extremely weak development of one of the six intergrowing individuals, but careful study of these complex interpenetration twins always establishes the presence of six individuals.

In transmitted light the disseminated cassiterite is reddish orange with unevenly distributed dark-brown spots and bands. In the cassiterite fourlings and sixlings there is a stronger dark-brown color in the centers of the intergrowth; this grades outward into brownish-orange.   There is weak zoning in the outer parts of the aggregates.

The relations between disseminated cassiterite and quartz in various parts of the vein are extremely contradictory. In the central part of the vein the disseminated cassiterite, forming both isolated crystals and chain and irregular accumulations, is always sharply idiomorphic toward quartz (Fig. 83).   Near the vein contacts the reverse is often the case; cassiterite here is usually segregated between quartz grains (Fig. 84).   Cassiterite is also idiomorphic toward fluorite, but stepped boundaries, corresponding to mutually grown faces, are sometimes observed between cassiterite and fluorite grains near the vein contacts. The relations between topaz and muscovite have been noted above.

The distinctive clotlike segregations of cassiterite which are present in the vein quartz in addition to uniformly disseminated cassiterite consist of fine-grained reddish-brown cassiterite. Along with the overall fine grain of the aggregates, in each clot there is a regular change of the size of the cassiterite grains inward. In the outer parts of the aggregates there are rather large (up to 1 mm) cassiterite crystals; toward the centers the grain size decreases sharply, and very fine-grained, weakly polarizing, sometimes completely isotropic aggregates usually predominate at the centers.

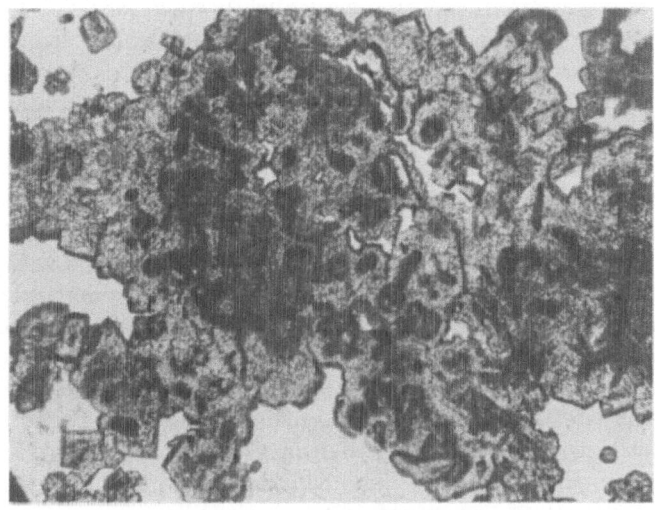

Fig. 85. Clotty segregations of fine-grained cassiterite with numerous relicts of cryptocrystalline cassiterite (× 70; uncrossed nicols).

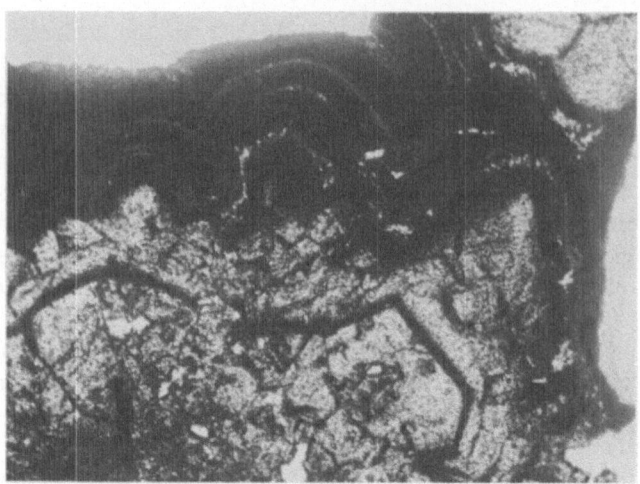

Fig. 86. Metacolloidal cassiterite at the margins of a clotty
segregation of cassiterite (x 90; uncrossed nicols).

Rounded segregations of dark-brown metacolloidal cassiterite, both
solitary and as irregular accumulations, are present in the fine-
grained regions (Fig. 85). Aggregates composed of very small, closely
packed spherulites are also sometimes present in these regions.

In the centers of the clotlike cassiterite segregations are great
numbers of very small cavities whose walls have more or less well-
defined reniform outlines. Also, relicts of spheroidal aggregations,
of which the entire aggregate probably consisted earlier, before it
lost its original appearance by recrystallization, are in a number of
cases clearly observed in these same regions (Fig. 85).

In some cases, thin fringes of dark-brown metacolloidal cassiterite
are present around the clotlike segregations. These show collomorphic
texture and distinct zoning caused by alternation of dark-brown and
yellowish-brown cassiterite. Cassiterite fringes have a well-defined
skeletal appearance. In the granular aggregates near these fringes
the cassiterite grains contain abundant inclusions of metacolloidal
cassiterite, mainly small and with regular outlines and, less commonly,
forming discontinuous zoned belts (Fig. 86). The presence of these
inclusions is unequivocal evidence that the granular aggregates formed
by recrystallization of metacolloidal cassiterite.

From the foregoing we see that in texture and morphology the
clotlike segregations of cassiterite are not uniform. These polytextural aggregates, characterized by the presence of several textural
varieties of cassiterite, are connected by intergradations. These

aggregates are formed in connection with recrystallization processes and are, so to speak, the connecting links between primary and secondary monotextural aggregates.

By monotextural aggregates the writer means mineral aggregates that preserve uniformity of texture in a given unit volume; for example, an aggregate has a single, rigorously determined texture— only collomorphic, only spherulitic, only granoblastic, etc. By polytextural aggregates the writer means mineral aggregates in which there are different kinds of textures interrelated by gradual transitions in a given unit volume; for example, collomorphic, spherulitic, and granoblastic textures are present together in an aggregate.

Table 13 gives the chemical composition of cassiterite forming clotlike segregations in quartz (analysis made by G. A. Arapova). The analysis shows a rather high tin content, which is explained by the presence of a mechanical admixture of wolframite in the cassiterite. The high iron content and the almost complete absence of manganese indicate more definitely that ferberite is present as a tin-bearing mineral admixture in the cassiterite. Data on the content of rare earths and scandium in the analysis can be referred completely to scandium, because no rare earth elements were detected spectrographically.

The following minor elements were identified spectrographically in the cassiterite: Fe, W (0.n); Sc, Ga (0.0n⁻); Mn, Ti (0.00n); Cu, In (0.00n⁻); and Be (0.000n). To study more precisely the nature of the minor elements in the cassiterite, the heavy fraction of the insoluble residue was analyzed spectrographically. The analysis

Table 13. Chemical Composition
of Cassiterite of the First Stage of
Mineralization

Component	%	Component	%
SiO_2	0.84	FeO	Not determined
Ti_2O	—	MnO	Trace
SnO_2	97.84	MgO	0.03
Al_2O_3	0.44	CaO	0.13
Fe_2O_3	0.70	WO_3	0.41
$\Sigma TR + Sc$	0.014		
		Sum	100.405

showed the presence of the following elements: Si (n); Al, Sn, W, Fe (n); Ti $(0.n^+)$; Na (0.n); Zr, Mg, Ca, Nb $(0.n^-)$; Ba $(0.0n^-)$; Mn, Cu (0.00n); and Cr, V, Ga, Be $(0.00n^-)$.

In comparing the results of spectrographic analysis of the cassiterite and the heavy fraction of the insoluble residue of the same cassiterite, the higher content of a number of minor elements, such as W, Fe, and Ti, and the appearance of other elements (Zr and Nb in substantial quantities, and Na, Cr, and Va in small quantities) was found.   This increase in the concentrations of these minor elements in the insoluble residue shows that these elements are present in minerals that are insoluble in HCl, namely zircon, ferberite, and ilmenorutile.   It is also possible that titanium and niobium are related, not to an ilmenorutile admixture, but to inclusions of rutile and mossite.

Wolframite.   Wolframite is found in isolated cases as dark-brown, thinly lamellar crystals sharply idiomorphic toward quartz. It is not found associated with other minerals.

Rutile.   Rutile, as very fine acicular crystals, solitary or forming paniculate growths, is also rare.

Thus, from the above we see that the mineral association in ores of the first stage of mineralization is characteristic of deposits of the cassiterite–quartz type.

In conclusion we should emphasize the following:

1. The predominant quartz and cassiterite are represented by crystallographic varieties.   Relicts of collomorphic texture and clear signs of recrystallization, expressed as nonuniform mosaic texture of grains, are noted in the quartz.   In contrast to the quartz, the cassiterite forms finer-grained aggregates, and, despite the obvious predominance of crystalline varieties, metacolloidal varieties are often also present in the cassiterite segregations.   Such a difference in the degree of the crystallinity of these minerals is explained by the different degree of the stability of the metacolloidal forms during metamorphism of the ores (Levitskii, 1953).

2. Morphologically the cassiterite crystals are characterized by short prismatic habit.   The forms $s\{111\}$, $m\{110\}$, $a\{100\}$, and $z\{321\}$ are observed visually.   The habit of the crystals is determined by the predominant development of $s\{111\}$ and $m\{110\}$.   Both the habit and the crystal forms determining it are characteristic of cassiterite from deposits of the cassiterite–quartz type (Gotman, 1941).

3. The mineral admixtures in the cassiterite of the first stage of mineralization are typical of cassiterite of this type.

4. The presence of 0.014% of scandium is characteristic of cassiterite of deposits of cassiterite–quartz type.

Thus, ores of the first stage of mineralization in the Shakh-Shagaila deposit display all the principal mineralogical features of deposits of the cassiterite–quartz type, from wall-rock alteration patterns to minor elements in cassiterite.

Mineralogy of Ores of the Second Stage of Mineralization

Ores of the second stage of mineralization are considerably less abundant in the deposit than ores of the first stage. They are restricted to the zone of intense brecciation of quartz–muscovite and topaz–muscovite greisens in the northern part of the greisen aureole of the first stage of mineralization and partly in granite unaffected by greisenization of the first stage.

The ores of the second stage are quartz–cassiterite veins cemented by greisen fragments and forming a complex network owing to numerous branchings and junctions among one another. At points of junction there are usually pocketlike bulges that are often 10 to 15 times thicker than the veinlets. Figure 87 is a sketch of the brecciated zone observed in trench 6b. It is seen from the figure that ores

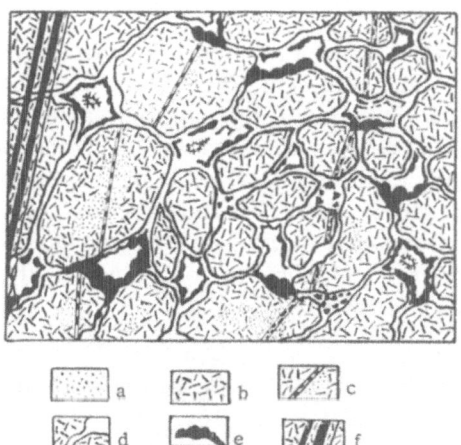

Fig. 87. Sketch of part of the brecciated zone of the second stage of mineralization. (a) Relict regions of quartz–topaz greisen; (b) completely muscovitized quartz–topaz greisen; (c) quartz–cassiterite veinlets of the first stage of mineralization; (d) quartz veinlets and geodes of the second stage of mineralization; (e) cassiterite in geodes; (f) adularia–hematite veinlets of the third stage of mineralization (× 5).

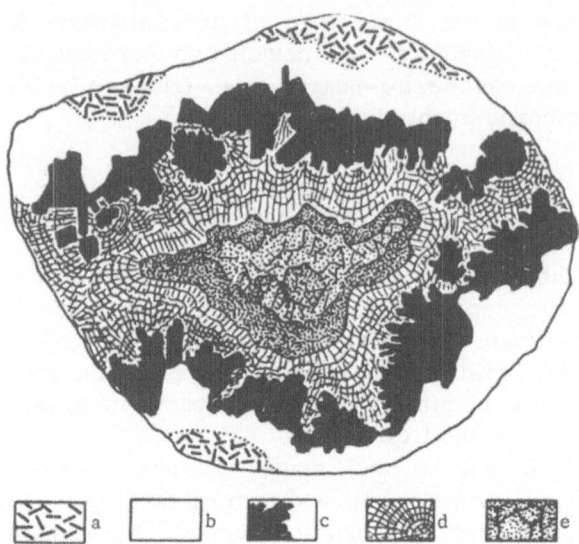

Fig. 88. Structure of mineralized cavity of the first type of mineralization. (a) Muscovitized topaz greisen; (b) zoned coarse-grained quartz; (c) cassiterite; (d) collomorphic medium-grained quartz; (e) drusy quartz (× 3).

of the second stage form a system of small mineralized cavities interconnected by thin veinlets. The thickness of the veinlets ranges from 4 to 10 mm. The size of the cavities varies from 2 by 3 cm to 7 by 12 cm. The veinlet contacts are very sharp. Vein quartz is in immediate contact with the wall rock and can be separated from it quite easily by slight impact. The contact surface of the vein filling is commonly an exact counterpart of all the irregularities of the fissure walls. The mineral composition of veins of the second stage of mineralization is as follows:

Predominant	Quartz
Very abundant	Cassiterite, fluorite
Moderately abundant	Adularia
Minor	Sericite
Rare	Topaz, pyrite, chalcopyrite

Thus, ores of the second stage are not distinguished by diversity of mineral associations, but their mineralogy is of considerable interest because of the great diversity of morphological varieties of cassiterite and textures of the mineral aggregates.

Wall-rock alteration is represented by quartz—muscovite greisen.

Structure of the Mineralized Cavities. As noted above, the main mineralization is restricted to the mineralized cavities of the small chamber type. Depending upon the proportions of the principal minerals, quartz and cassiterite, and the predominance of one or another morphological variety of cassiterite, the following types of mineralization of the cavities are recognized:

1. Cavities filled mainly by medium-grained quartz with distinctly collomorphic texture. Cassiterite is present in the quartz as a discontinuous zone of spherulitic aggregates close to the boundaries of the cavity (Fig. 88). Immediately at the contact with the wall rock there is always a thin fringe of coarse-grained zoned quartz. Quartz is the predominant mineral; the ratio of quartz to cassiterite in this

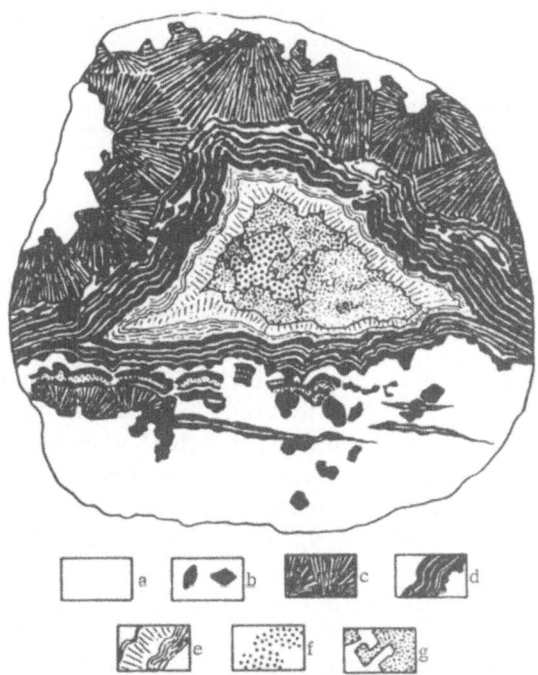

Fig. 89. Structure of mineralized cavity of the second type of mineralization. (a) Coarse-grained quartz; (b) coarse-grained cassiterite; (c) finely acicular spherulitic cassiterite; (d) metacolloidal cassiterite; (e) collomorphic medium-grained quartz; (f) dense, finely lamellar sericite; (g) drusy quartz (× 3).

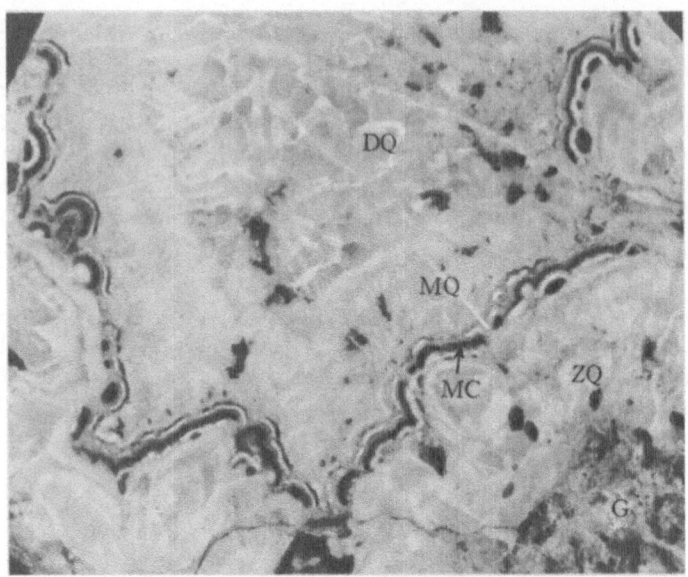

Fig. 90. Structure of mineralized cavity of the third type of mineralization.
(G) Quartz-muscovite greisen; (ZQ) coarse-grained zoned quartz; (MC)
metacolloidal cassiterite; (MQ) medium-grained quartz; (DQ) drusy
quartz. The black dissemination is coarsely crystalline cassiterite (x 3).

type of mineralization is 10:1 on the average. Generally there are
small cavities with drusy quartz and fluorite in the central parts of
the filling.

2. Cavities filled with fine-grained collomorphic quartz and cassiterite. In many cases cassiterite predominates over quartz (Fig. 89).

The cassiterite is of three morphological varieties: (a) spherulitic—forming a continuous zone of closely packed spherulites; (b)
collomorphic, cryptocrystalline—forming a series of variously colored,
festooned zones overlying the spherulitic zone; (c) earthy, reniform
—filling the central voids.

The distribution of minerals in fillings of this type is as follows:
Along the contact with quartz—muscovite greisen there is a thin
zone of coarse-grained zoned quartz, followed inward by a thicker
zone of spherulitic cassiterite. This in turn gives way inward to
cryptocrystalline collomorphic quartz; and finally the center of the
cavity is filled by earthy reniform cassiterite in some cases and by
sericite and fluorite in other cases.

3. Cavities filled mainly by quartz of several varieties with cassiterite sharply subordinate (Fig. 90). At the walls of the cavity is a

zone of coarse-grained, comby, zoned quartz. Rare segregations of coarse-grained cassiterite, also zoned, are observed between the quartz grains.   This zone is up to 1.5 cm thick.   Inward there is a zone of metacolloidal cassiterite.   The thickness of the cassiterite zone is nowhere greater than 2 mm, but within it the metacolloidal cassiterite, segregated as globules, globulites, oolites, spherulites, and microreniform, hemispherical, and festooned aggregates, shows great morphological variety. The next zone inward consists of medium-grained quartz with relicts of collomorphic texture.   The quartz contains sparsely disseminated microspherulites and fragments of festooned cassiterite. The center is filled with coarse-grained quartz, with rose-colored adularia between the grains. Sometimes a rather thick drusy cavity filled with fluorite is present at the center.

It is seen from this description of the types of mineralization of the cavities that, despite certain differences in mineralization the following general structural patterns are evident:

1. The presence of a zone of coarse-grained zoned quartz always at the margins of the cavity.
2. The restriction of fine-grained and in some cases also collomorphic aggregates of vein quartz and metacolloidal aggregates of cassiterite to the intermediate zones of the fillings.
3. The universal presence of small drusy cavities in the centers of the fillings.

Passing to a description of the minerals, the diversity of textural and morphological varieties of vein quartz, which despite the predominance of coarse-grained aggregates can be considered metacolloidal quartz, deserves particular emphasis.

Texture and Morphology of the Vein Quartz

The quartz that fills the cavities forms predominantly medium-grained and coarse-grained aggregates, the greater part preserving some metacolloidal features.   Strictly metacolloidal quartz, as fine-grained aggregates closely associated with metacolloidal cassiterite, is also present.

As already noted, in all three types of mineralization of the cavities there is a strict regularity in the distribution of the main varieties of vein quartz. The sequence of zones from the contacts to the centers of the cavities is: (1) coarse-grained or comby zoned quartz; (2) metacolloidal or fine-grained quartz; (3) medium-grained quartz with relicts of collomorphic texture; (4) coarse-grained drusy quartz in the center of the filling.

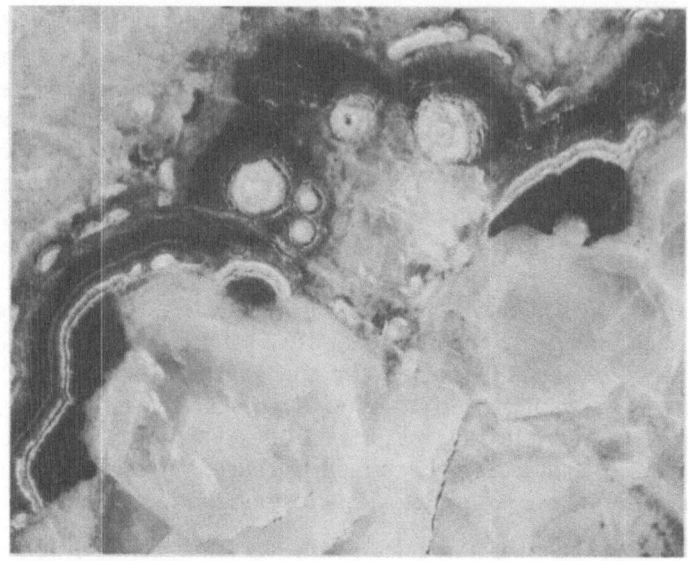

Fig. 91. Coarse-grained zoned quartz from the marginal zone of a
mineralized cavity of the third type. Black, gray, and white festooned
aggregates are metacolloidal cassiterite. Relict regions of metacolloidal
quartz and cassiterite can be seen in the aggregate of coarse-grained
quartz in the right-hand part of the photograph (x 15).

Coarse-Grained Comby Quartz. The coarse-grained
comby quartz forming the outer zones of mineralized cavities of all
types shows distinct zoning. The center of the crystal is light-gray
transparent quartz with clear crystallographic outlines. This is the
core of the zoned crystal. The size of these cores usually does
not exceed 2 to 3 mm and only rarely reaches 5 mm. These homogeneous transparent cores are surrounded by envelopes of milky
translucent quartz (Fig. 91) with a characteristic zoning caused by
the alternation of thin zones of grayish translucent quartz with zones
of milky porcelanous quartz. The width of these zones ranges up
to 0.1 mm. The number of zones varies with the location of the
crystal. Their number is usually greater in zoned crystals near the
contact with the zone of metacolloidal cassiterite.

It is seen under the microscope that the central cores of the
crystals show uniform extinction in some cases and relict sectorial
extinction in other cases. Near the boundaries of the cavities, the
cores of the zoned crystals are more uniform. Toward the zone of
metacolloidal cassiterite, sectorial extinction, evidence of non-
homogeneous rosette structure, is more common in the cores. A

similar pattern is noted in the distribution of gas and liquid inclusions in the cores. In crystals near the boundaries of the cavities, the cores contain small numbers of inclusions, but cores of zoned quartz crystals near the metacolloidal cassiterite zone contain considerable numbers of unordered gas and liquid inclusions.

The boundaries between the cores and the zoned envelopes are indistinct and gradual. In most cases the boundaries are distinctively dendritic and denticulated.

Both the overall outline of the envelopes and the outlines of each of the constituent zones repeat the crystallographic outlines of the cores. The presence of extremely large numbers of gas and liquid inclusions forming series of radial chains is very characteristic of the zoned envelopes.

The various zones differ from one another in the texture of the quartz forming the envelope. Zones of thinly columnar, sometimes fibrous quartz alternate with zones of coarsely columnar quartz. As there are no parting surfaces between zones, the zones cannot be considered to be the result of interrupted growth of the crystals with successive deposition of material on the faces of earlier crystals. Zones with finely columnar texture can be considered zones of skeletal growth; Fig. 92 is representative of these. In the photograph, which includes part of the core of a zoned quartz crystal and a few zones of the envelope, it is clearly seen that many finely columnar quartz

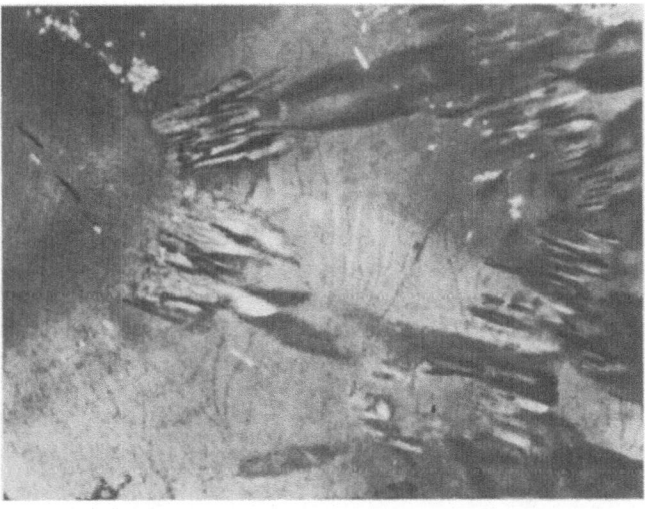

Fig. 92. Radial-block texture of zoned quartz (x 70; crossed nicols).

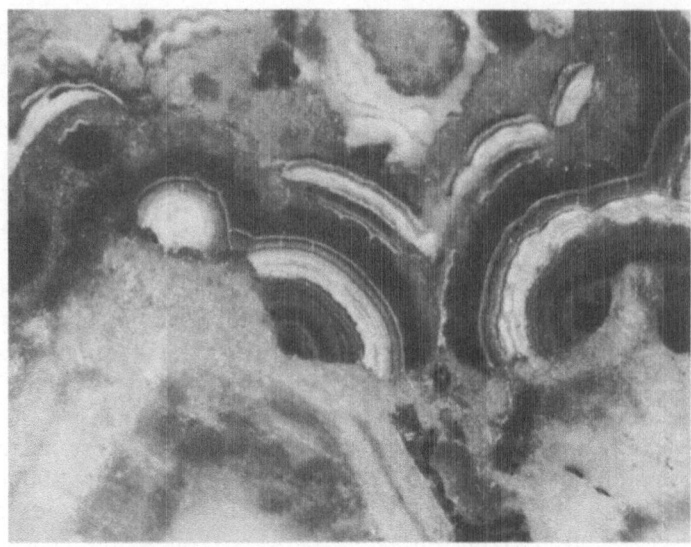

Fig. 93. Relations between coarse-grained zoned quartz and metacolloidal
cassiterite (x 20).

crystals are skeletal forms of growth of the central core and of the
larger crystals in zones consisting of coarsely columnar quartz.

The nonuniform extinction of the individual blocks is related to
the small change in orientation (of the order of several degrees) of
the "secondary" blocks filling the space between the main skeletal
blocks.   Such a disorientation of the individual blocks is mainly
caused by anomalous conditions of crystal growth, particularly
skeletal growth. A very small mechanical admixture of metacolloidal
cassiterite in the metacolloidal quartz grains, caught during skeletal
crystallization, disturbs the regular growth of the individual blocks,
leading to a small deviation of the crystallizing blocks from the
crystallographic orientation of the "main" crystal and to the formation
of boundary surfaces between these blocks.

Thus, the quartz crystals forming the outer zones of the mineralized cavities have radially mosaic structure formed during skeletal
growth of these crystals. Zoning of the crystals is caused by periodic
changes in the conditions of crystallization, which determine whether
there will be normal or skeletal growth of individual crystals.

The appearance of zoning in the quartz crystals forming the
coarse-grained aggregates is varied.   Zoning is almost absent in
crystals immediately adjacent to the boundaries of the cavity, but
5 to 7 mm from the boundaries all crystals are zoned, without excep-

tion. There are transitional areas in which the zoning of crystals is distinctly asymmetrical—the width of the envelopes surrounding the homogeneous cores is sharply greater toward the zone of meta-colloidal cassiterite. Finally, close to the cassiterite zone the quartz crystals show more or less uniform development of zoned envelopes. Relict regions of fine-grained metacolloidal quartz and metacolloidal cassiterite are sometimes observed in the coarse-grained quartz mass (see Fig. 91). At the boundary with the metacolloidal cas-siterite zone the number of relict regions of metacolloidal quartz is markedly greater.

Coarse-grained zoned quartz is closely associated with coarse-grained cassiterite that is unevenly distributed in the quartz mass. Relations between zoned quartz and cassiterite are highly contra-dictory; in some cases quartz is sharply idiomorphic toward cas-siterite, but in other cases the reverse is clearly observed. Stepped mutually grown boundaries, evidence of concurrent growth of quartz and cassiterite, are most common.

The relations between zoned quartz and metacolloidal cassiterite are of great interest. Figure 93 clearly shows that crystallization of the zoned quartz was accompanied by intense solution and recrystal-lization of metacolloidal cassiterite and partial replacement by colum-nar quartz of the outer envelopes. At the same time, festooned aggre-gates of metacolloidal cassiterite underwent some mechanical deformation, expressed as fractures in the festooned and spherical aggregates. The characteristic feature of these disturbances is, on the one hand, that they appear only in the metacolloidal cassiterite aggregates and do not extend into neighboring zones and, on the other hand, that they are present everywhere in connection with the pene-tration of aggregates of zoned quartz into the metacolloidal cassiterite region. Thus, there is reason to suppose that these fractures of the festooned aggregates are the result of crystallization pressure developed during crystallization of the zoned quartz.

In conclusion we give the results of a semiquantitative spectro-graphic analysis of the transparent quartz of the central cores and the milky quartz of the envelopes. The former is very pure. The following elements are found in it besides silicon: Na $(0.0n^{-})$; Fe, Mg, Sn $(0.00n)$; Ca, Cu, Al $(0.00n^{-})$; and W, Be, and Mn (traces of lines). The number of admixtures in the zoned quartz is much greater. In addition to silicon the following elements were identified in the milky quartz: Sn $(0.n^{-})$; Na $(0.0n^{+})$; W, Fe $(0.0n^{-})$; Y, Mn, Sb, Cu $(0.00n)$; Mg, Ca, Al, Ti $(0.00n^{-})$; and Bi and In (traces of lines). The increased content of Sn, W, Fe, and a number of other elements is caused by finely dispersed relict admixtures of cassiterite and

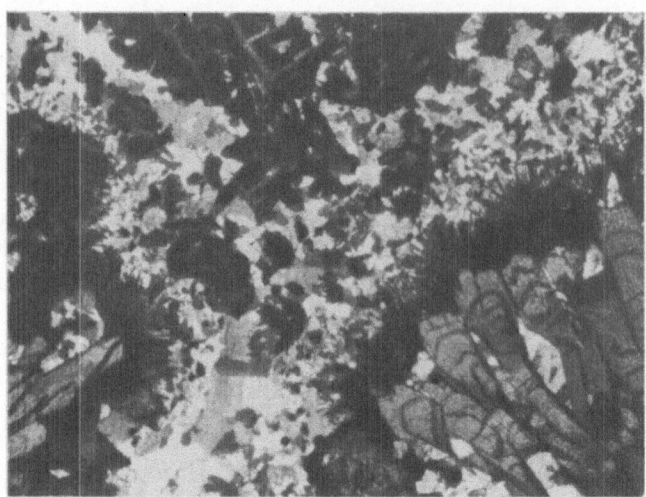

Fig. 94. Metacolloidal quartz zone in mineralized cavities of the second type. Black, metacolloidal cassiterite; gray, acicular cassiterite; lighter fine-grained aggregate, metacolloidal quartz (x 46; crossed nicols).

wolframite that were enclosed by the quartz in replacement of the metacolloidal quartz—cassiterite aggregates during recrystallization.

Metacolloidal Quartz. Metacolloidal quartz is closely associated with metacolloidal cassiterite. It is moderately abundant in the second and third types of mineralization of the cavities. In the first type of mineralization, it is noted only as relict regions in medium-grained quartz containing relicts of collomorphic texture. It shows distinct recrystallization features. It is present as fine-grained aggregates with mosaic texture, in which small spheroidal aggregates of metacolloidal cassiterite are often present.

In the second type of mineralization, metacolloidal quartz forms small zones interlayered with festooned aggregates of metacolloidal cassiterite, and in a number of cases it forms a zone that separates the zones of spherulitic and metacolloidal cassiterite (Fig. 94). It forms fine-grained aggregates with granoblastic texture of the main mass and minor polytextural character expressed in the appearance, in the main even-grained mass, of irregular and sometimes linear regions of coarser-grained quartz with radially mosaic structure of the grains. Such regions can be considered centers of recrystallization.

Metacolloidal quartz is also closely associated with metacolloidal cassiterite in cavities of the third type of mineralization. In contrast to the second type of mineralization, the metacolloidal quartz here does not form continuous zones but is present only as isolated accumulations restricted to regions surrounded on almost all sides by dense, festooned aggregates of metacolloidal cassiterite.   Only in these regions was the original shape and texture of the cassiterite and quartz aggregates preserved: globular cassiterite aggregates, quartz–cassiterite aggregates with characteristic gel texture (see Fig. 102), almost isotropic quartz, and fine-grained aggregates of metacolloidal quartz (Fig. 95).   Even in these regions, centers of recrystallization, which are represented by groups of enlarged grains with prismatic outlines and uneven, denticulated boundaries and which have radially mosaic structure (noticeable, however, only at high magnifications), are noted in the metacolloidal quartz aggregates.   Metacolloidal quartz also forms the inner zones of reniform and complex spherical aggregates of metacolloidal cassiterite (Fig. 96).

Thus, taking into account (1) the presence of relict aggregates of metacolloidal quartz, together with metacolloidal cassiterite, in the coarse-grained zoned quartz mass, (2) the presence of series of isolated regions of metacolloidal quartz in the inner zones of the festooned aggregates of metacolloidal cassiterite, and (3) the universal presence of metacolloidal quartz within reniform and spherical aggregates of metacolloidal cassiterite, it can be hypothesized that metacolloidal quartz occupied the preexisting volume but during subse-

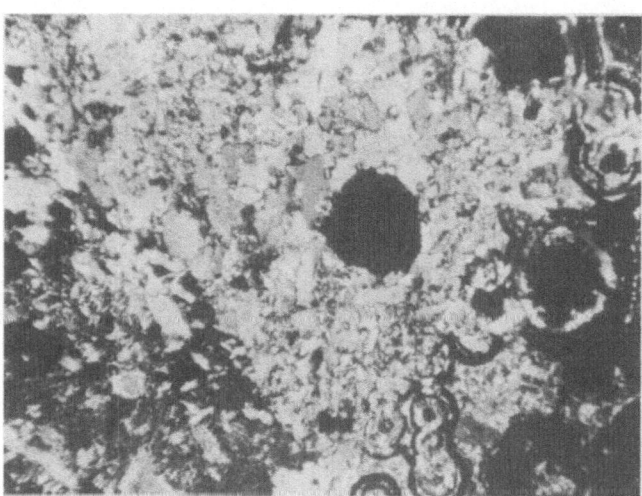

Fig. 95. Metacolloidal quartz within a cassiterite zone. Mineralized cavities of the third type (x 70; crossed nicols).

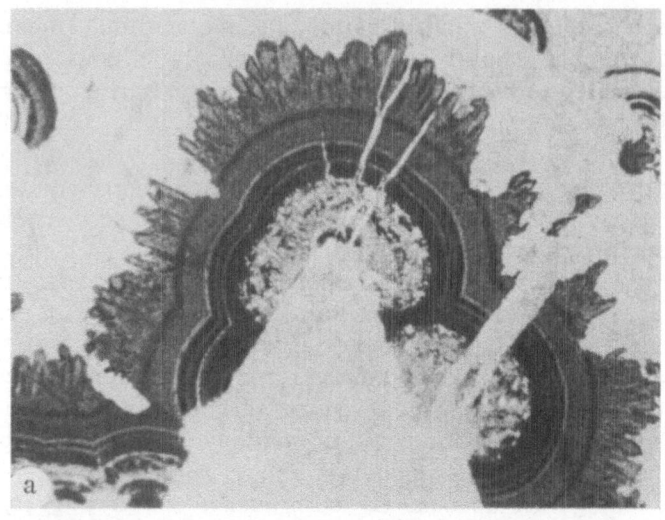

Fig. 96. Metacolloidal quartz in the inner zones of reniform aggregates of metacolloidal cassiterite (× 46). (a) Crossed nicols; (b) uncrossed nicols.

quent recrystallization was almost completely replaced by coarse-grained quartz. The metacolloidal quartz was preserved only where it was most isolated by dense crusts of metacolloidal cassiterite.

Medium-Grained Quartz. Medium-grained quartz is the dominant variety in cavities of the first and third types of mineralization. Medium-grained quartz is rare in cavities of the second type mineralization. The characteristic feature of this quartz is the universal presence in it of aggregates with relict collomorphic texture. These are readily observable in thin sections with uncrossed nicols; with crossed nicols the relicts of collomorphic texture are lost in the overall hypidiomorphic-granular quartz mass and are distinguished with great difficulty. In some cases no features of collomorphic texture are observable in thin section either with or without crossed nicols.

In a polished section representing the mirror image of a given thin section, however, clear features of collomorphic texture can be observed at low angles of illumination (Fig. 97). This is especially characteristic of medium-grained quartz in mineralized cavities of the first type. Spherical aggregates composed of quartz or of quartz and cassiterite are observed in this type of cavity in isolated instances. The quartz forming these aggregates is fine-grained and medium-grained in some cases and rather coarse-grained in others. Irrespective of composition, all these spherical aggregates are closely associated with cryptocrystalline cassiterite that forms irregular accumulations in the medium-grained quartz. Small spheroids composed of two or three quartz grains forming complex skeletal intergrowths are observed in this cassiterite.

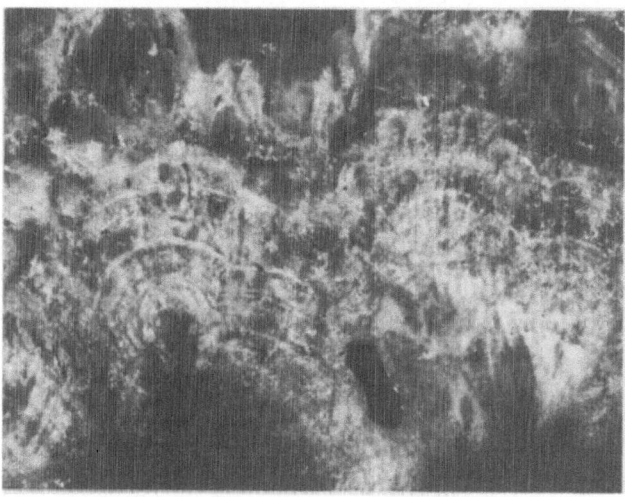

Fig. 97. Relicts of collomorphic texture in aggregates of medium-grained quartz (x 20).

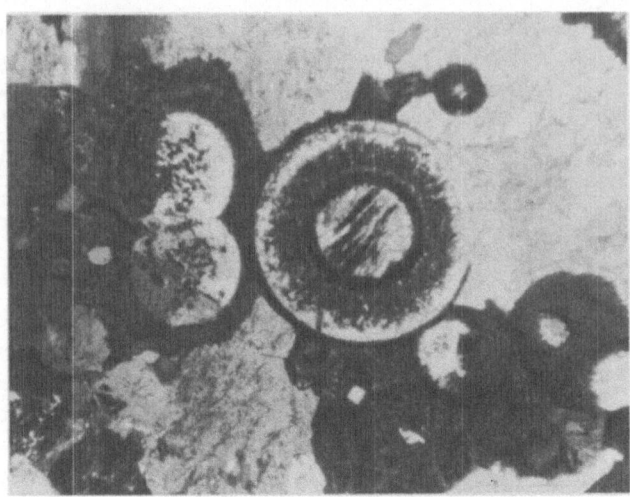

Fig. 98. Spheroidal aggregates of quartz and quartz—cassiterite composition. In the center of the photograph, quartz of the marginal zone of the large spheroid is seen to blend with the quartz ground-mass (x 46; crossed nicols).

In many cases there is crude zoning in the quartz—cassiterite spherical aggregates. The centers of these aggregates are composed of fibrous quartz in some cases and of rather coarse-grained quartz aggregates in others (Fig. 98). Away from the center of the aggregate is a zone of cryptocrystalline cassiterite. Usually this zone is quite thin, but in some cases the cryptocrystalline cassiterite forming the zone occupies up to 50% of the spheroid. The outermost zone consists of the fine-grained and medium-grained quartz. The boundaries of this quartz with the cassiterite are extremely uneven and bear the character of complex intergrowths. When the spherical aggregates are not overlain by cryptocrystalline cassiterite, the quartz of the outermost zone merges with the medium-grained quartz of the groundmass. In this, individual quartz grains are common both to aggregates of the surrounding even-grained groundmass and to aggregates forming the outermost zone of the spheroid (Fig. 98).

Coarsely spheroidal aggregates of medium-grained quartz con-sisting of several rather large grains (usually two to three times coarser than the groundmass grains) with radially mosaic internal structure are more common. These aggregates contain inclusions of fine cassiterite oolites or quartz—cassiterite ooids whose size does not exceed that of the oolites, ranging up to 0.1 mm.

The coarsely spheroidal granular aggregates are usually bordered by plumose or comby quartz, which at the margins grade into the even-grained groundmass (Fig. 99). There is reason to suppose that these are recrystallized exfoliation spheroids.

Though the medium-grained quartz groundmass is even grained on the whole, the texture shows a certain porphyritic character. This is caused by recrystallization and is observed in regions where the recrystallization was not complete. The characteristic feature of such regions is the presence of isolated, rather-coarse quartz grains, with the interspaces filled with finer-grained quartz and, in a number of cases, also with metacolloidal cassiterite. The coarse crystals everywhere show features of skeletal growth. Their boundaries with the fine-grained aggregates show the development of a complex system of skeletally branching growth forms (Fig. 100). Solution of meta-colloidal cassiterite oolites along the skeletal-crystallization front of the quartz, with subsequent deposition of acicular cassiterite crystals on the opposite surfaces of the oolites is clearly observable in these regions.

During skeletal growth of two neighboring grains, there is a characteristic pattern of mutual intergrowth of two skeletal systems. During forward growth these systems of crystal branches move toward one another and, by coming into contact, form complex inter-lacings and intergrowths. In this, despite certain distortions, each skeletally branching system retains the orientation of the host crystal.

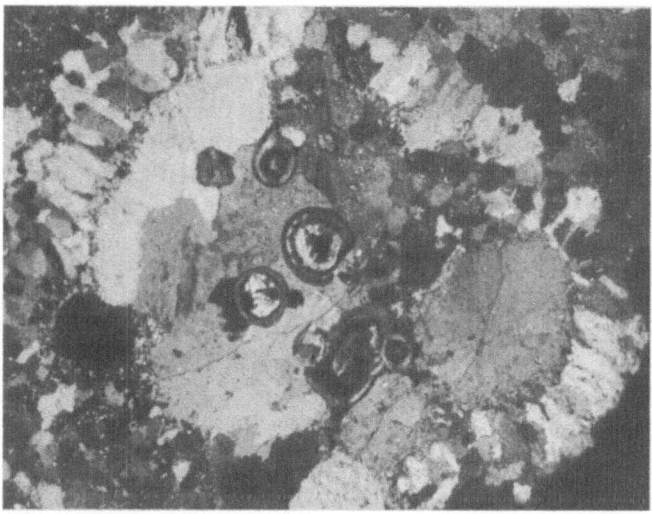

Fig. 99. Coarsely spheroidal aggregate of medium-grained quartz. Small spherical quartz–cassiterite aggregates are seen in the center (× 46; crossed nicols).

Fig. 100. Juncture of two differently oriented skeletal systems, inheriting the optical orientation of the central core (x 70; crossed nicols).

In regions of medium-grained quartz in which recrystallization was most complete, distinct skeletal-growth textures are rare, and most of the grains preserve only relicts of skeletal-mosaic structure. There are grains in which even relicts of this structure are absent, and only rare, radial chains of gas and liquid inclusions mark the weakly distinguishable traces of skeletal growth.

It should be noted that metacolloidal cassiterite associated with medium-grained quartz and forming part of the finely festooned and complex spherical aggregates was also affected by recrystallization.

Thus, aggregates of medium-grained quartz forming the intermediate regions of the mineralized cavities show well-defined polytextural character caused by varying intensity of recrystallization.

On the basis of what has been described above, a number of distinguishing features of recrystallization in the vein-quartz aggregates can be recognized. One of the most general features of recrystallization is the polytextural character of the aggregates. The intensity of recrystallization can be judged by the degree of polytextural character of the mineral aggregates, determined by the relative proportions of primary, transitional, and secondary textures. Other characteristic features are: (1) more or less distinct forms of skeletal growth of the grains; (2) mosaic structure of the grains, caused by skeletal crystallization (showing radially mosaic or block-mosaic internal structure of the grains, depending upon how the grain is cut); (3) relict inclusions of metacolloidal mineral aggregates that were recrystallized; and (4) relict traces of microstructures and textures of the original aggregates. Identification of these features in the mineral aggregates permits us to settle unambiguously the problem of the occurrence and intensity of recrystallization.

Drusy Quartz. Drusy quartz is observed in small quantities in the mineralized cavities. Generally it almost completely fills the centers of the mineralized cavities. Freely growing crystals, represented by combinations of a prism $m\{1010\}$ and two equally developed rhombohedra $r\{1011\}$ and $z\{0111\}$, are sometimes present. The faces of these forms are covered by numerous vicinal faces. The quartz is predominantly milky, less often grayish and translucent. A small number of gas and liquid inclusions arranged in zones can be seen under the microscope. The following minor elements were found in the drusy quartz spectrographically: Ca, Al $(0.0n^+)$; Na $(0.0n)$; Sn, Fe $(0.0n^-)$; Mg, Y $(0.00n^+)$; Ti, W, Mn, Cu $(0.00n)$; Sb, Bi $(0.00n^-)$; and In (traces of lines).

Thus, the vein quartz that fills veinlets and cavities of the second stage of mineralization shows unmistakable features of colloidal origin, which are masked somewhat by recrystallization.

Analysis of the texture and morphology of the vein-quartz varieties described above and of the interrelationships of these varieties shows that the main mass of the quartz filling the cavities was originally metacolloidal quartz, which later was recrystallized. This is evidenced by relict regions of metacolloidal quartz and traces of original collomorphic texture in medium-grained quartz aggregates (intermediate zones) which were more or less recrystallized. Similar relict aggregates of metacolloidal quartz associated with metacolloidal cassiterite are also observed in the coarse-grained zoned quartz forming the outer zones of the mineralized cavities. In this, metacolloidal quartz and cassiterite are often present as inclusions directly in the crystals of this quartz.

The writer pictures the process of mineralization and subsequent formation of the vein fillings of cavities as the following:

1. The cavities were filled by a complex $SiO_2 + SnO_2$ gel. While still in a viscous state, this gel underwent a number of transformations: (a) primary segregation into layers with the formation of isolated monomineralic masses (Levitskii, 1953); (b) compaction and decrease in volume of the overall gel mass of the filling and, in connection with this, the exfoliation of the gel mass from the cavity walls;

(c) the development of a central void by continuing compaction and decrease in volume of the gel mass; and (d) the squeezing of the solution into the central void and into the exfoliation cavities.

2. The siliceous gel crystallized to form a fine-grained aggregate of metacolloidal quartz. At the same time, quartz crystallized on the cavity walls from true solutions squeezed into exfoliation cracks upon decrease in volume of the gel mass. This might have led to the overall crystallization of the gel. During this, the origin and growth of larger crystals in the immediate proximity of the fine-grained metacolloidal aggregates (which might not yet have had the consistency of a solid) could have been the impulse for recrystallization of the primary metacolloidal quartz aggregates. The decrease in silica concentration in the solution, owing to the growth of quartz crystals on the cavity walls, brought these solutions to extreme unsaturation with respect to the very fine metacolloidal quartz grains and thereby caused them to be resolved.

Thus, the crystallization front of the quartz near the cavity boundaries can be at the same time a front of solution of metacolloidal quartz. It is characteristic that the growth of large crystals and solution of fine-grained aggregates were not concurrent. A decrease in the concentration of the solution did not at once caused solution metacolloidal quartz; only a decrease in concentration to the very lowest level led to intense solution of metacolloidal quartz and thus to strong supersaturation of the solution by large crystals. This latter condition corresponds to rapid skeletal growth and inclusion of a great quantity of the mother liquor, with the formation of radial gas and liquid inclusions. With decreasing concentration there was normal crystallization until attainment of the lowest concentration level that could cause renewed solution of metacolloidal quartz and renewed supersaturation of the solution by large crystals. This cycle was repeated many times, leading ultimately to coarse-grained aggregates of zoned quartz; the zoning reflects this cyclicity.

The metacolloidal quartz aggregates forming the inner zones of the mineralized cavities also did not retain their original appearance. Recrystallization went on here too, but with a lower intensity that varied from region to region in these zones. The unevenness of recrystallization caused the variety in textures of these quartz aggregates, in which, in a small area relicts of collomorphic, metacolloidal, granoblastic, crystalloblastic, porphyritic, and pseudographic textures are observed under the microscope. Such polytextural character is considered one of the most characteristic features of recrystallization of mineral aggregates.

Texture, Morphology, and Composition
of Metacolloidal Cassiterite

After quartz, metacolloidal cassiterite is the most widespread mineral in ores of the second stage of mineralization and displays great diversity of morphological varieties. In mineralized cavities of the first type metacolloidal cassiterite is present rarely as irregular accumulations in quartz and less often as concentric zones in quartz–cassiterite aggregates. Metacolloidal cassiterite is in many cases the dominant mineral in cavities of the second type. It is present as finely festooned aggregates interlayered with collomorphic quartz. Metacolloidal cassiterite, forming a thin zone referred to above as the metacolloidal cassiterite zone (see Fig. 90), is present in small quantities in mineralized cavities of the third type.

Because the metacolloidal cassiterite aggregates in mineralized cavities of the three types differ substantially in the degree of crystallization, their textural-morphological varieties are divided into two subgroups:

1. Primary, aggregates which are almost unaffected by recrystallization
2. Secondary, aggregates which formed by recrystallization of the primary metacolloidal aggregates

Primary Textural-Morphological Varieties of Metacolloidal Cassiterite. The primary textural-morphological varieties of metacolloidal cassiterite show extremely weak anisotropy and sometimes even complete isotropy under the microscope and are represented by specific forms characteristic of colloidal aggregates.

The greatest diversity of morphological varieties of metacolloidal cassiterite of this subgroup is in mineralized cavities of the third type. Globules, aggregates close to globulites, oolites, complex spheroidal reniform aggregates, and discontinuous festooned and reniform aggregates are present. Figure 101 graphically illustrates the structure of the mineralized cavities and the distribution of most of the varieties of metacolloidal cassiterite listed above.

As previously noted, a zone of comby quartz, consisting of a coarse-grained aggregate of zoned crystals in which there are relict regions of fine-grained metacolloidal quartz, is directly adjacent to the cavity walls. Metacolloidal cassiterite is present here only as relics. Rare, relatively coarse cassiterite crystals, closely asso-

Fig. 101. Structure of part of a mineralized cavity of the
third type. Explanation in text.

ciated with coarse-grained zoned quartz, were formed by recrystallization of metacolloidal cassiterite.

The next zone inward in the cavity consists of metacolloidal
cassiterite. The zone is inhomogeneous in structure; in different
parts it consists of different morphological types of metacolloidal
cassiterite. Reniform and hemispherical forms overlain by festooned
collomorphic aggregates are predominant. The collomorphic aggregates are generally fringed by a dense crust of finely acicular cassiterite.

At the boundary of this zone are regions of fine-grained metacolloidal quartz containing globular segregations of metacolloidal
cassiterite. Globular cassiterite aggregates similar to globulites
and also cassiterite oolites are present here as well. The lower part

of Fig. 101 illustrates such a region. The quartz—cassiterite zone, next inward, is composed of medium-grained quartz containing regularly oriented fragments of reniform accumulations of cassiterite oolites and spherulites. Relicts of collomorphic texture in the quartz are quite often seen under the microscope. Clear features of recrystallization of the quartz—cassiterite collomorphic aggregates, with the formation of spherulites of cassiterite and granoblastic aggregates of medium-grained quartz, are noted within these zones.

The centers of the mineralized cavities of this type are in most cases completely filled by coarse-grained drusy quartz. Less commonly, small cavities filled with well-formed crystals of quartz, fluorite, and acicular cassiterite are present.

Globules. Globular forms of metacolloidal cassiterite are rare and are restricted exclusively to relict regions of metacolloidal quartz. It was noted earlier that metacolloidal quartz is observed at the boundary between the zones of metacolloidal quartz and coarse-grained zoned quartz, and it must be emphasized that this quartz was preserved only within deep reentrants of the dense metacolloidal cassiterite crust. As a rule, the metacolloidal quartz in such reentrants consists of fine-grained aggregates with granoblastic or crystalloblastic texture, but often in the deepest parts of the reentrants there are cryptocrystalline cassiterite—quartz masses. These show

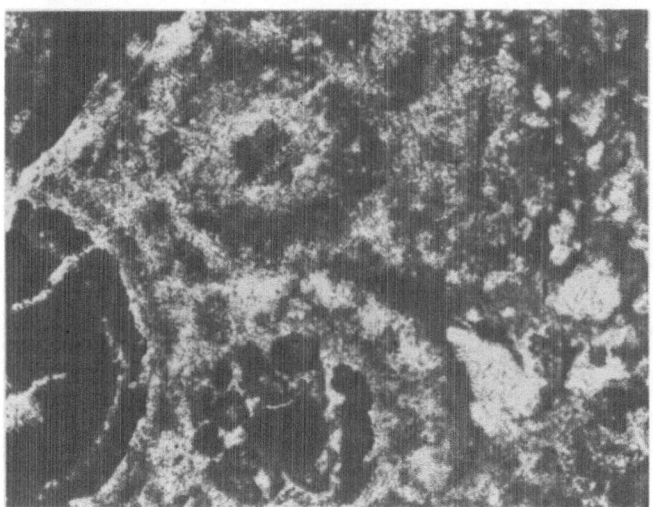

Fig. 102. Clotty spheroidal texture of cryptocrystalline cassiterite-quartz aggregate (x 150).

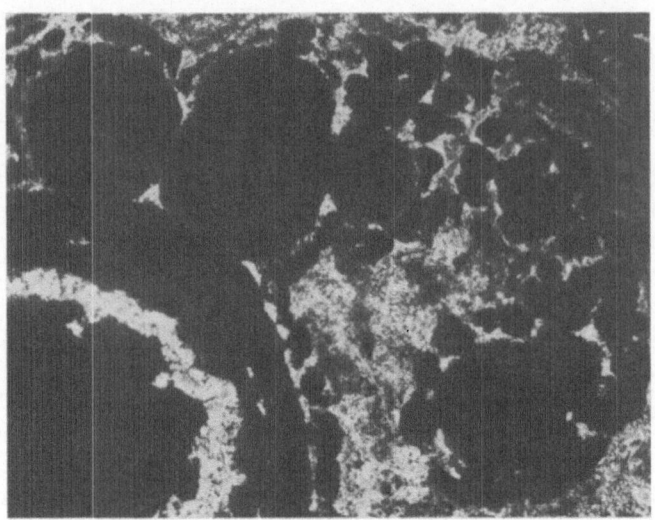

Fig. 103. Globular segregations of metacolloidal cassiterite in cryptocrystalline quartz. In the lower right corner is a spherical aggregate of globulite type (x 150; uncrossed nicols).

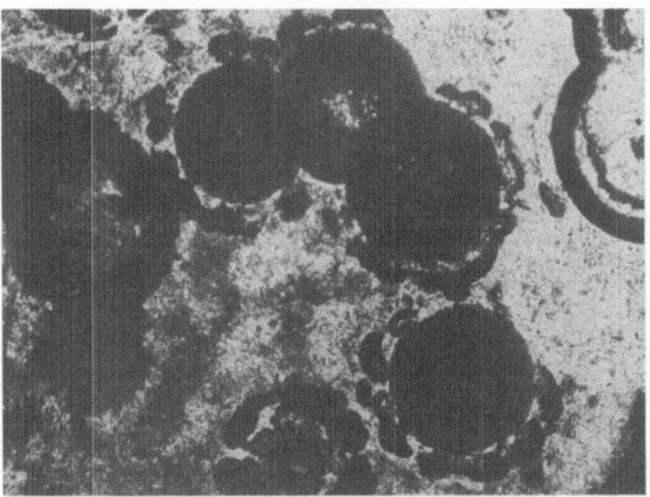

Fig. 104. Complex spheroidal aggregates of metacolloidal cassiterite of globulite type in cryptocrystalline cassiterite–quartz aggregate (x 150; uncrossed nicols).

a distinctive clotlike spheroidal texture that reflects the primary stages of segregation of the two-component $(SiO_2 + SnO_2)$ gel into layers (Fig. 102).

As seen in the photograph, very fine flakes of metacolloidal cassiterite grouped around certain centers representing denser clots of this cassiterite are regularly distributed in the cryptocrystalline quartz groundmass. The impression is that the flaky particles of colloidal SnO_2 clumped together to form globules. The writer directly observed such clumping of very fine flakes to form globules during synthesis of spherical aggregates of metacolloidal sphalerite; usually, immediately after coagulation of the zinc sulfide gel the structureless mass of the gel, consisting of very fine flaky particles, separated into numerous clumps, at first friable, but after some time becoming more distinctly spherical. Thus, the clotlike spheroidal texture of the cryptocrystalline quartz—cassiterite aggregates reflects the initial stage of diagenetic alteration of the two-component $SiO_2 + SnO_2$ gel.

The distribution of cassiterite globules in the cryptocrystalline quartz is quite uniform, but in some cases group segregations of globules are observed (Fig. 103). Among group accumulations, coa-

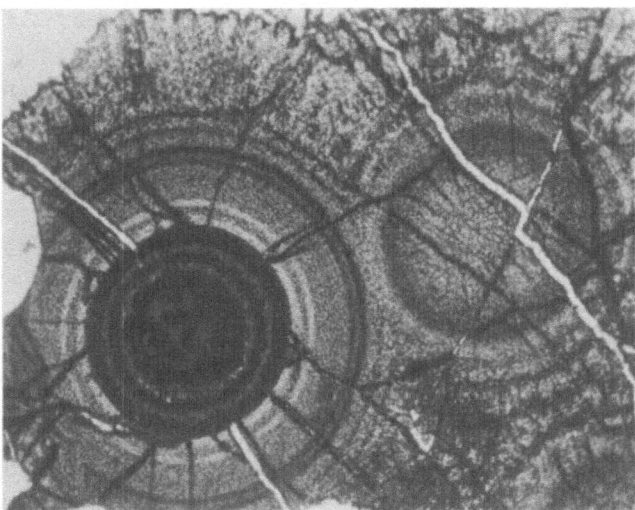

Fig. 105. Quartz—cassiterite oolites incrusted by finely acicular cassiterite. In the center ot the large oolite is a globular core from which radial veinlets of dark metacolloidal cassiterite branch (x 70; uncrossed nicols).

Fig. 106. Diffusional zoning of an oolite composed of cryptocrystalline quartz—cassiterite aggregate (x 70).

lescent globules and aggregations of numerous closely packed globules
are often observed. Such aggregations may be considered transitional
to globulites (coarser spherical aggregates with globular structure).
The size of the globules is 0.005 to 0.15 mm.

Globulites. Larger spherical cassiterite aggregates of globulite
type are observed in these same regions of cryptocrystalline quartz
in addition to globules. These are aggregates of separated or partially
coalescent globules (Fig. 104). Sometimes the groups of concentric
globular chains alternate with cryptocrystalline quartz. At high
magnifications a nonuniform structure of the large spheres can be
distinguished. The greater part of their volume consists of coalescent,
rather large globules. The size of these somewhat anomalous
globulites ranges up to 0.3 mm.

Under the microscope with crossed nicols the cassiterite of the
forms described above is completely isotropic. Only in isolated
cases in the centers of the coarse globulites are there regions of
irregular shape consisting of fine-grained cassiterite.

Oolites. In addition to globules and globulites, in regions of cryptocrystalline quartz there are also cassiterite oolites with clear concentrically zoned structure.

The zoning is caused by rhythmic alternation of cassiterite and
quartz zones. The centers of the oolites are always composed of
small globulites. The cassiterite of the oolites is isotropic, as is the
cassiterite of the globulites.

Cryptocrystalline cassiterite oolites are most common in aggregates of medium-grained quartz forming the intermediate zone of the mineralized cavity.

The great majority of the oolites are more or less regularly spheroidal and have surfaces covered by a thin, discontinuous crust of crystalline-granular cassiterite. Some oolites, especially those in regions of most intense recrystallization, are surrounded by halo-shaped aggregates of finely acicular cassiterite (Fig. 105).

The oolites described above show typical diffusional zoning. The boundaries between differently colored shells are blurred and gradual; there are no parting surfaces between shells (Fig. 106). Zoning of the oolites is caused by diffusional crystallization of pigmenting components.

The texture of the quartz—cassiterite aggregates forming the oolites is metacolloidal granular. The texture is uniform over the entire aggregate, and does not change from one shell to another. There are dark cores with globular texture in the centers of many oolites. Oolites whose outer zones are cut by very fine radial veinlets of dark cassiterite branching off from the central cores of the oolites are common (Fig. 105). Generally these veinlets do not extend outside the oolites but instead wedge out in the outer shells. Sometimes they penetrate into the halo-shaped aggregates of finely acicular cassiterite

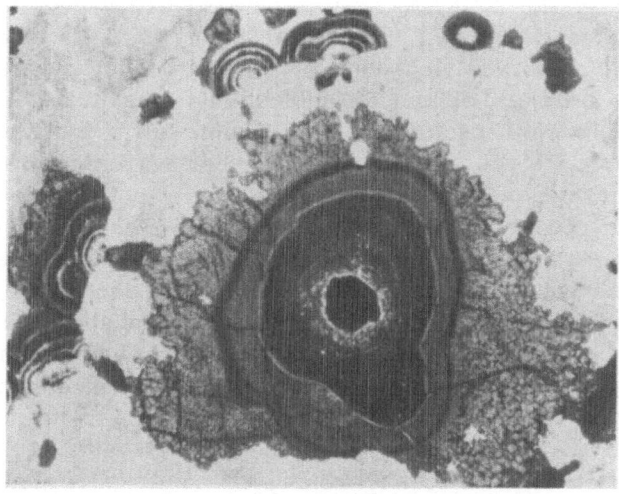

Fig. 107. Irregularly spheroidal aggregate of metacolloidal cassiterite incrusted by finely acicular cassiterite. Globular metacolloidal cassiterite is at the center of the aggregate (× 70; uncrossed nicols).

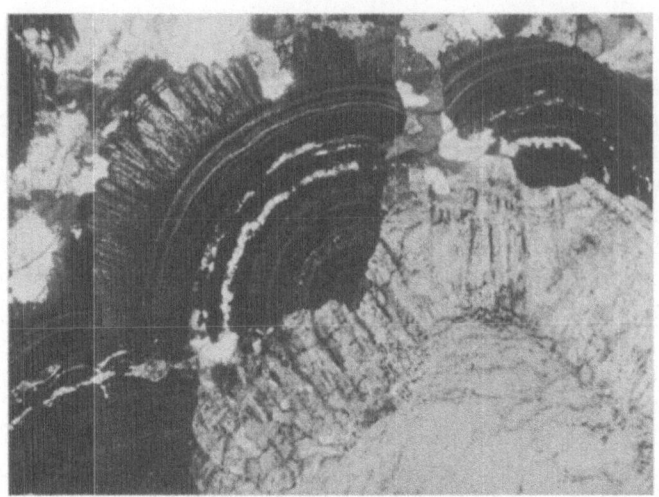

Fig. 108. Relations between hemispheroidal aggregates of metacolloidal cassiterite and zoned quartz (x 46; crossed nicols).

and into neighboring oolites when group aggregations of oolites are
present.   The formation of such veinlets is evidence that the inner
parts of the oolites remained in the gel condition for a long time while
the main mass of the oolites (and perhaps all the cavity filling as well)
was consolidated.

Distinctive irregularly spheroidal forms of metacolloidal cassiterite are observed in the medium-grained quartz aggregates in
addition to ordinary oolites of metacolloidal cassiterite (Fig. 107).
Like the oolites, these forms have globular cores and typical diffusional zoning.   Such irregularly spherical aggregates can be considered deformed oolites.

As noted above, the continuous cassiterite zone is somewhat nonuniform in structure.   It consists of separated hemispheres overlain
by a continuous reniform crust.   On the surface of the reniform
aggregates is a crust of acicular cassiterite crystals. The reniform
aggregates of metacolloidal cassiterite form an irregular, closed
zone roughly parallel to the cavity walls.   The lower part of the
continuous cassiterite zone consists of hemispherical cassiterite
aggregates.   At the boundary with the comby quartz, the individual
hemispheres show concentric zoning in some cases and globular
texture in others.   At their bases the hemispheres form stepwise
curved contact surfaces with the comby quartz crystals, caused by
the deforming effect of growing quartz crystals on the still relatively

viscous metacolloidal cassiterite (Fig. 108). Because the coarse-grained aggregates of zoned quartz were formed during recrystallization of metacolloidal cassiterite, some of the hemispherical and globular aggregates of metacolloidal cassiterite were also affected by recrystallization. Regions of crystalline-granular cassiterite, usually retaining relict collomorphic zoning, developed in the centers of these hemispheres (Fig. 109).

The hemispherical aggregates in turn hindered normal growth of quartz—it is often observed in thin sections that, when crystals are surrounded by large hemispheres, they usually are skeletal forms. The separated hemispherical aggregates and the isolated globulites are overlain by reniform aggregates (Fig. 101).

Reniform aggregates. In cross section the reniform aggregates of metacolloidal cassiterite show a well-defined concentrically laminated structure, caused in most cases by repeated alternation of differently colored cassiterite. An alternation of cassiterite with fine-grained metacolloidal quartz is less common (Fig. 110).

As a rule, the inner zones are composed of darker varieties of metacolloidal cassiterite, and the outer zones, of lighter varieties. In the outer zones there are numerous dehydration cracks along which dark cassiterite penetrates from the inner zones; this is evidence

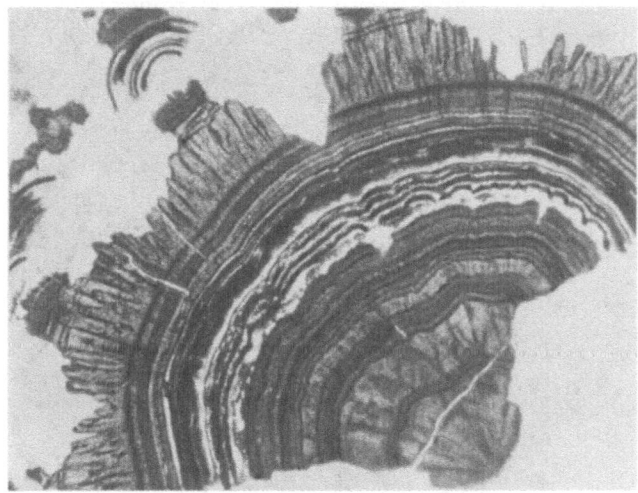

Fig. 109. Initial stages of recrystallization of metacolloidal cassiterite in hemispherical aggregates of this cassiterite. In the center: a single-crystal core with relicts of metacolloidal SnO_2 (× 70; uncrossed nicols).

Fig. 110. Reniform aggregates of metacolloidal cassiterite whose inner
zones are enriched in metacolloidal quartz.

of the later consolidation of the cassiterite gel in these zones. The
cassiterite forming the reniform aggregates is mainly isotropic, but
in the outer zones it is noticeably anisotropic.

The following elements were identified spectrographically in the
cryptocrystalline reniform cassiterite from mineralized cavities of
the third type: Fe $\cong$ 1%; Al, W (0.0n$^+$); Na, Mn, Sb (0.0n); Mg (0.0n$^-$);

Table 14. Chemical Composition of Reniform Cassiterite of the Second Stage of Mineralization

Component	%	Component	%
SnO_2	90.16	WO_3	0.21
SiO_2	7.14	FeO	Not determined
TiO_2	0.02	MnO	<0.001
ZrO_2	0.01	MgO	None
Al_2O_3	0.40	CaO	0.12
Fe_2O_3	0.94	Sb	0.18
TR + Y	0.02	Cu	0.023
		$H_2O^\pm$	0.50
		Sum	99.724

Ga, Cu $(0.00n^+)$; Ca, Pb, Bi $(0.00n)$; Ti, In $(0.00n^-)$; V $(0.000n)$; Ag, Be $(0.000n^-)$.

Contraction cavities of exfoliation, with reniform aggregates of metacolloidal cassiterite on the walls, are often observed among the dense cryptocrystalline quartz–cassiterite aggregates filling mineralized cavities of the second type. The surfaces of the hemispheroidal elements are even but slightly rough. The roughness is caused by numerous, very small (0.1 mm) discoidal aggregations of chalcedony unevenly distributed on the surfaces.

Table 14 gives a chemical analysis of this cassiterite, made by A. T. Sadikova. The analysis shows an elevated silica content caused by an admixture of metacolloidal quartz forming very fine zones interlayered with metacolloidal cassiterite. The appreciable iron content is probably caused by an admixture of finely dispersed hematite and ferberite. A substantial antimony content and the presence of copper is characteristic. The data on content of rare earths and yttrium can be almost entirely referred to yttrium, because no rare earth elements were determined spectrographically and the content of scandium, similar in properties, ranges from 0.001 to 0.003%.

Besides the admixtures determined by chemical analysis, the following minor elements were identified spectrographically in this

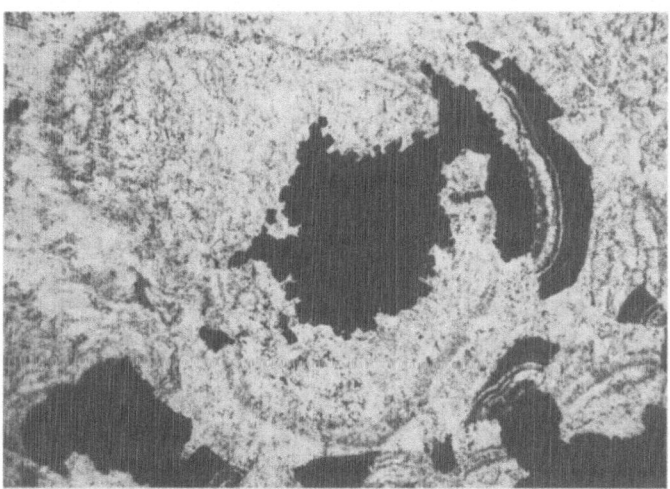

Fig. 111. Recrystallized reniform-spheroidal aggregate of metacolloidal cassiterite. The marginal parts are almost completely replaced by quartz; the original outlines of the aggregate are marked by relict inclusions of metacolloidal cassiterite (x 70).

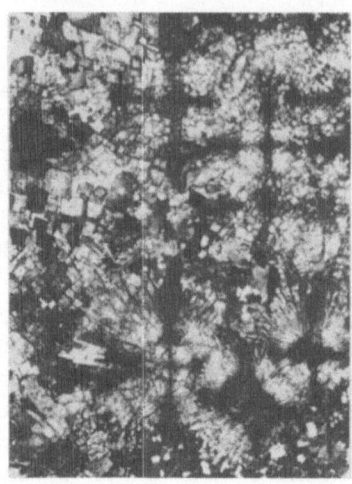

Fig. 112. Granular aggregate of cassiterite with relicts of spherulitic
texture (x 46; crossed nicols).

cassiterite: Be (0.0n); Ga (0.0n$^-$); Bi, In (0.00n); Sc (0.00n$^-$); V
(0.000n$^-$).

Secondary Textural-Morphological Varieties of
Metacolloidal Cassiterite. As noted above, the primary
quartz—cassiterite metacolloidal aggregates filling the mineralized
cavities were appreciably recrystallized. Metacolloidal quartz, whose
recrystallization features were described above, was affected most
strongly. Metacolloidal cassiterite shows much greater stability
during recrystallization, but it too was appreciably recrystallized.
In addition to oolites of metacolloidal cassiterite, complex reniform-
spheroidal aggregates of metacolloidal cassiterite are observed in
the medium-grained quartz forming the intermediate zone in mineralized cavities of the third type. The parts of these aggregates in regions
of most intense recrystallization of the medium-grained quartz were
also almost entirely recrystallized. As a rule, the complex reniform-
spheroidal aggregates at the periphery consist of dense metacolloidal
cassiterite sometimes interlayered with fine-grained quartz. The
centers are composed of fine-grained quartz. During recrystallization of these aggregates, the outer zones of metacolloidal cassiterite
were replaced by quartz and one or more spherulites of acicular
cassiterite formed in their centers. Distinct outlines of the replaced
aggregates are preserved in the pseudomorphic quartz owing to very
fine inclusions of relict metacolloidal cassiterite (Fig. 111). In regions
of most intense recrystallization of quartz and the complex reniform-
spheroidal quartz—cassiterite aggregates, metacolloidal cassiterite
oolites were also considerably recrystallized to form typical sphe-

rulites. Depending upon the intensity of recrystallization, in some cases relict oolitic or globular texture is preserved in the centers of the spherulites and, in other cases, the spherulites grade into granular aggregates with characteristic extinction crosses (Fig. 112).

In mineralized cavities of the second type the densely packed spherulites form a continuous zone between the zones of coarse-grained zoned quartz and collomorphic cryptocrystalline cassiterite. The relations between the spherulitic cassiterite aggregates and the cryptocrystalline varieties of cassiterite are evidence that the spherulitic aggregates are secondary forms that developed during recrystallization of cryptocrystalline collomorphic cassiterite.

By studying the relations between the acicular cassiterite crystals forming the spherulites and the isotropic metacolloidal cassiterite overlying the spherulitic aggregates, it was established that the thickness of the metacolloidal cassiterite crust varies sharply from one spherulite to another. Also, its thickness often varies substantially even over a small segment in contact with a single spherulite. If the spherulitic aggregates formed first and the metacolloidal cassiterite aggregates formed subsequently, the latter would uniformly coat the surfaces of the former and the thickness of the coating would be the same everywhere. Thus, the observed relations between the spherulitic cassiterite aggregates and the metacolloidal cassiterite aggregates cannot be considered the result of progressive precipitation of initially crystalline and later metacolloidal cassiterite from solution; on the contrary, the spherulitic aggregates are later, secondary forms arising during recrystallization of metacolloidal cassiterite.

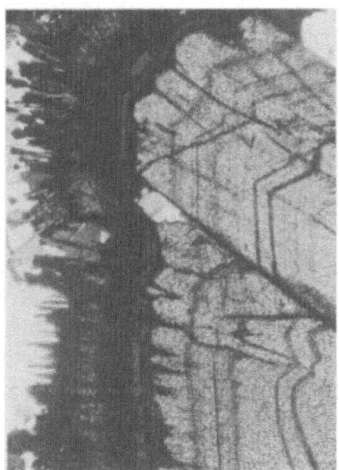

Fig. 113. Skeletal appearance of cassiterite crystals in contact with metacolloidal cassiterite (x 70; crossed nicols).

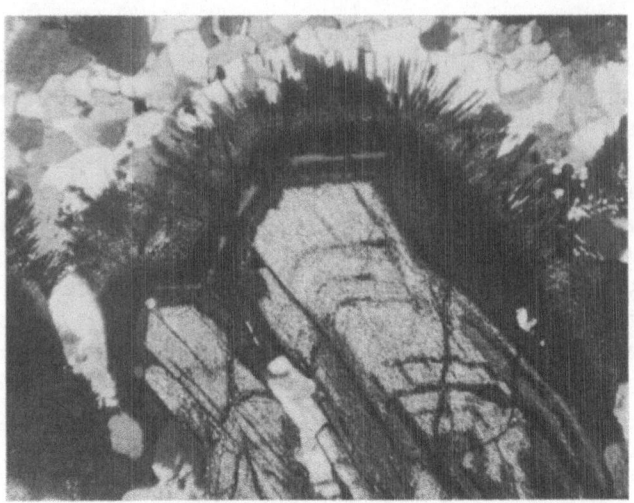

Fig. 114. Relict zoning of cassiterite crystals caused by inclusions of metacolloidal cassiterite. Light gray, crystalline cassiterite; dark gray and black, metacolloidal cassiterite; granular aggregate in upper part of photograph, quartz (x 70; crossed nicols).

There are numerous secondary structural features in the crystals themselves forming the spherulites. Among these are:

1. The skeletal aspect of the cassiterite crystals at their boundaries with the metacolloidal cassiterite (Fig. 113).

2. The presence in the crystals of inclusions of metacolloidal cassiterite, in some cases arranged in zones (Fig. 114) and in other cases forming variously oriented linear segregations at the boundaries of the numerous sectors of the crystals (Figs. 113 and 116).

3. The mosaic structure of the cassiterite crystals forming the spherulitic aggregates (Fig. 115). This is caused by skeletal growth of these crystals and conditions unfavorable for crystallization, probably increased viscosity of the medium.

It is characteristic that there is always a zone of weakly polarizing metacolloidal cassiterite at boundaries between the crystals forming the spherulites and the isotropic zones of metacolloidal cassiterite (Fig. 116). The nonuniform structure of this zone becomes apparent at high magnifications; very fine cassiterite crystals are more or less uniformly distributed in the main isotropic mass of metacolloidal cassiterite. The number of these very small crystals increases toward the spherulite crystal, and they acquire an orientation the same as that of the spherulite crystal.

Also, some of the crystals forming the spherulitic aggregates are distinctive crystals composed of very fine shafts whose interspaces are filled with metacolloidal cassiterite. Such forms are the skeletal framework, so to speak, of the crystals and can be considered transitional between the usual "monolithic" crystals with mosaic texture and accumulations of very small, separated crystals in anisotropic metacolloidal cassiterite.

It must be noted that the writer (Lebedev, 1954a) observed similar skeletal-framework crystals of galenite in metacolloidal sphalerite in the metacolloidal lead—zinc ores of Iokun'zh (Tadzhik SSR) and Olkusz (Poland). Posěpný (1874) noted the same kind of galenite crystals in the Raibl lead—zinc deposits in Italy.

The formation of skeletal-framework crystals is apparently a general feature of crystallization of minerals in viscous media. Some experimental work also confirms this feature. Mokievskii (Mokievskii and Semenyuk, 1962) showed by experiments on the growth of alum and sodium chloride crystals that skeletal forms of the crystals of these compounds develop in viscous media. The present writer has observed very small skeletal galenite crystals that formed in a synthetic two-component gel of lead and zinc sulfides.

The foregoing facts undoubtedly attest to the secondary nature of the cassiterite spherulites. In thoroughgoing recrystallization, metacolloidal cassiterite is completely recrystallized into spherulitic aggregates and less commonly into fine-grained aggregates surrounding the cassiterite spherulites (Fig. 117).

Fig. 115. Block structure of cassiterite crystal (× 200; crossed nicols).

Fig. 116. Zone of anisotropic metacolloidal cassiterite at the contact
between cassiterite crystals and isotropic varieties of metacolloidal
cassiterite (x 90; uncrossed nicols).

Table 15 gives the chemical composition of the dense spherulitic
aggregates of cassiterite (analysis by G. A. Arapova). The results of
this analysis show that during recrystallization the kinds of admixtures in the cassiterite do not change. There is only some quantitative
decrease of them, particularly of silica.

In mineralized cavities of the first type, where metacolloidal
cassiterite is not present, zoning caused by inclusions of metacolloidal
cassiterite is very common in cassiterite crystals (Fig. 118).

Table 15.   Chemical Composition of
Spherulitic Cassiterite of the Second
Stage of Mineralization

Component	%	Component	%
SnO$_2$	97.59	WO$_3$	0.39
SiO$_2$	0.88	FeO	Not determined
TiO$_2$	Not determined	MnO	Trace
ZrO$_2$	" "	MgO	0.05
Al$_2$O$_3$	0.86	CaO	0.13
Fe$_2$O$_3$	0.60	SbO	0.14
TR + Y	0.015	Cu	Not determined
		Sum	100.655

In summing up, it is noteworthy that cassiterite of the second stage of mineralization shows characteristic features of colloidal origin, as does the vein quartz of this stage. One of the primary features is the presence of globular forms of metacolloidal cassiterite in regions of cryptocrystalline quartz. Also noteworthy is the different behavior of quartz and cassiterite during recrystallization. Whereas the original appearance of the quartz aggregates is altered completely by this recrystallization and only extremely rare relict regions of metacolloidal quartz or traces of collomorphic texture remain in the crystalline-granular aggregates, almost the reverse is the case for cassiterite. Though in a number of cases the cassiterite aggregates also lose their original appearance by recrystallization, primary textural-morphological varieties of cassiterite are dominant. This agrees well with the data of Levitskii (1953), who noted that cassiterite is substantially inert during recrystallization and metamorphism.

From the chemical and spectrographic analyses above it is seen that cassiterite of the second stage of mineralization, characterized by substantial admixtures of Y and Cu and the universal presence of Sb, Bi, and In (thousandths of a per cent), differs substantially from cassiterite of the first stage of mineralization in the character of the minor elements.

Appreciable quantities of fluorite, adularia, pyrite, and iron-rich muscovite and also extremely rare chalcopyrite are present in ores

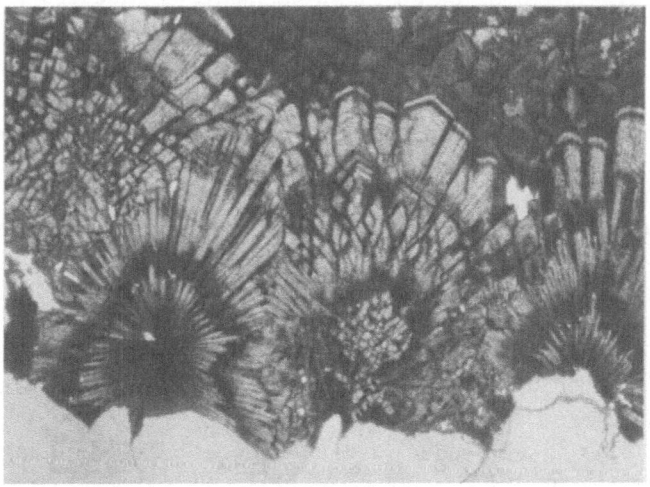

Fig. 117. Spherulitic cassiterite aggregates. In the upper part of the photograph is a region of completely recrystallized metacolloidal cassiterite (x 46; uncrossed nicols).

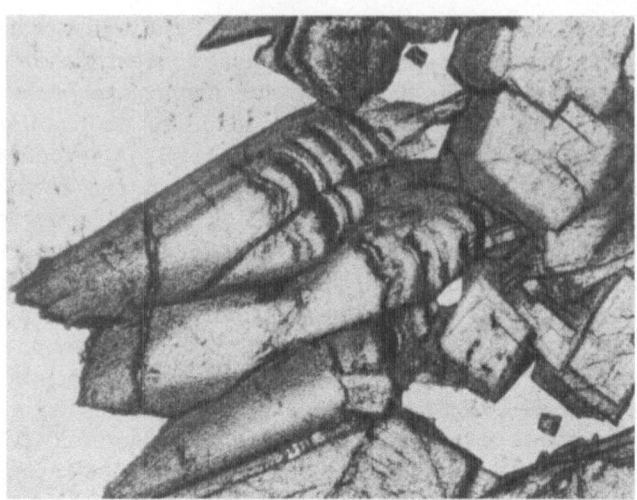

Fig. 118. Zoning in cassiterite crystals caused by inclusions of
metacolloidal cassiterite (x 70; uncrossed nicols).

of the second stage of mineralization in addition to the quartz and
cassiterite described above.

Fluorite. Fluorite is present in mineralized cavities of all types
but is abundant only in mineralized cavities of the third type, where it
completely fills spaces between quartz crystals in the center of the
cavity. It is light violet, with unevenly distributed dark-violet
regions. Rare, irregular gas and liquid inclusions are present; the
volume of the gas phase in these inclusions does not exceed 5 to 7%
of the total volume of the inclusions.

In addition to the main components, the following minor elements
were identified in the fluorite spectrographically: Sn, Al $(0.0n^+)$; Si,
Mg $(0.0n)$; Zn $(0.0n^-)$; Fe $(0.00n^+)$; Mn $(0.00n^-)$.

Adularia. Adularia is abundant in mineralized cavities of the
first and third types, both as isolated wedge-shaped crystals and as
irregular accumulations in the vein quartz. During recrystallization
of the vein quartz the greater part of the adularia was replaced by
quartz. The adularia is light-rose colored. The indices of refraction
are: $\alpha = 1.520 \pm 0.002$; $\beta = 1.523$; $\gamma = 1.524 \pm 0.002$; $2V$ measured on a
universal stage is 58 to 60°.

The following elements were identified in the adularia spectrographically: Si (n^+); K, Al (n); Na $(0.n)$; Fe $(0.n^-)$; Sn $(0.0n^+)$; Mg, Ca,
Ba $(0.0n^+)$; Y $(0.0n^-)$; and Mn and Ga $(0.00n^-)$.

Chalcopyrite.   Chalcopyrite is found as isolated, very small segregations in dense spherulitic aggregates of cassiterite in mineralized cavities of the first type.

Muscovite and pyrite, which are found in greisens of the second stage, were briefly described above.

In conclusion, it is noteworthy that that ores of the second stage of mineralization are transitional between ores of the quartz–cassiterite and cassiterite–sulfide types, both in mineral associations and in the nature of the minor elements in the cassiterite.

Mineralogy of Ores of the Third Stage of Mineralization

Ores of the third stage of mineralization are a rather thick series of convergent and coarsely parallel quartz, adularia–hematite, and pyrite veinlets, with northeast-southwest strike and characterized by the presence of great numbers of thin apophyses connecting series of parallel veins, which gives rise to a similarity with the latticework zones. The mineral composition of the veins is as follows:

Strongly predominant	Quartz, adularia, hematite
Very abundant	Pyrite, chlorite, fluorite
Moderately abundant	Calcite, chalcopyrite
Minor	Cassiterite
Rare	Galenite

The surrounding granite is chloritized in the vein zones. The quartz–muscovite greisens in the eastern limb of the brecciated ore zone of the second stage of mineralization and the topaz–muscovite and quartz–muscovite greisens of the latticework zones of the first stage of mineralization were intensely chloritized.

Granite wall-rock alteration in the vein zones of the third stage of mineralization has been described in more detail on page 107.

As already noted, veins of the third stage of mineralization may be subdivided compositionally into adularia–hematite, quartz, and pyrite veins.   Mineralization of these veins took place at different times.   This is evidenced both by the mineral associations and by the observed superposition and, less commonly, truncation of earlier veins by later veins.

A decrease in the concentration of oxygen in the solution with a corresponding increase in hydrogen sulfide concentration is a general feature in the change of mineral associations from earlier to later veins.

Adularia–Hematite Veins. Adularia–hematite veins, 1 to 8 mm thick, are the earliest veins. Though macroscopically distinct, the vein contacts are diffuse under the microscope owing to gradual transition from wall rock to vein.

Hematite. The zones near the vein contacts consist of crystalline hematite forming mostly continuous borders. Discontinuous fringes in which hematite forms isolated radiating aggregations consisting of coarser lamellar crystals are noted as well. Sometimes isolated spherulites are present in the central parts of adularia veins. The hematite is sharply idiomorphic toward adularia and chlorite.

Adularia. Adularia, forming yellowish-rose to brownish-red fine-grained aggregates, is the principal vein mineral. The color distribution is uneven. At high magnifications, accumulations of finely crystalline hematite are noted in the dark varieties of adularia. It was established spectrographically that the iron content in the light varieties of adularia ranges from $0.0n^+$ to $0.n\%$; the iron content is always greater than 1% in the dark varieties.

Irregular accumulations of finely lamellar, dark-green chlorite are noted in the adularia close to the hematite fringes near the vein contacts.

Quartz Veins. The next youngest veins are quartz veins, whose mineral composition is more varied. Fluorite, chlorite, and less commonly calcite, cassiterite, and chalcopyrite are abundant here in addition to quartz. The veins are 1 mm to 3 cm thick, and the contacts are uneven and slightly blurred.

Quartz. The quartz forming the veins is light gray and fine-grained. Under the microscope it shows even-grained texture. The grain boundaries are wavy. The extinction is uniform. Gas and liquid inclusions are not abundant. Tabular inclusions with unordered distribution in the grains predominate. Accumulations of dustlike inclusions of finely crystalline cassiterite and very small chlorite flakes are present here and there. The following elements were found in this quartz spectrographically: Al, Fe $(0.n^-)$; Na, Mg, Cu $(0.0n)$; Ca, Sn $(0.0n^-)$; Ga, V $(0.00n)$; and Pb, Y, and Sn (traces of lines).

Fluorite. Fluorite usually forms irregular segregations in the quartz. In places it completely fills the veins for short distances. Euhedral crystals are less common. The fluorite is colorless, less commonly light violet. It contains abundant gas and liquid inclusions characterized by negligible volume of the gas phase and extremely uneven distribution in the crystals. Fluorite is usually closely associated with chlorite in the veins. The following elements were found in the fluorite spectrographically: Si $(0.n^-)$; Fe, Al $(0.0n)$; Na, Mg, Y $(0.0n^-)$; Sn, Cu, C $(0.00n^+)$; and Mn and In (traces of lines).

Chlorite.   Chlorite is found as vermicular, finely segmented inclusions forming irregular accumulations in quartz and fluorite. The chlorite is dark green, less often bottle green. The mean index of refraction is 1.665. The following elements were found spectrographically in chlorite separated from fluorite: Si, Fe (n^+); Al (n); Mg (n^-); Na, Mn $(0.n^+)$; Ca $(0.n)$; V $(0.0n)$; Ga $(0.00n)$; Sn, Cu, Y $(0.00n)$; and Ti (traces of lines). Both the mean index of refraction and the spectrographic data show that this chlorite is an iron-rich and magnesium-poor variety of aphrosiderite (daphnite) type, which is very characteristic of tin-ore deposits of the cassiterite–sulfide type.

Cassiterite.   Cassiterite is present as very small yellowish-gray acicular crystals unevenly distributed in the quartz. The forms $m\{110\}$ and $s\{111\}$ are noted. The cassiterite is sharply idiomorphic toward quartz and fluorite; in some cases it forms close intergrowths with chlorite lamellae, and in other cases it is sharply idiomorphic toward chlorite.   It is not found associated with chalcopyrite. The following elements were found spectrographically in this cassiterite: Sn (n^+); Fe (n^-); Si $(0.n)$; Na, W, In $(0.n^-)$; Mg, Cu $(0.0n^+)$; Mn $(0.0n^-)$; Al, Ga $(0.00n^+)$; and Ca, Pb, and Ti $(0.00n)$.

It is seen from this that the minor elements in cassiterite of the third stage of mineralization differ substantially from those in cassiterite of the first and second stages. The content of iron and copper increases substantially, a considerable admixture of manganese appears, and the indium content increases sharply. The tin content is nearly the same, and such elements as Nb, Sc, Y, and Sb are entirely absent.   This association of minor elements is typical of cassiterite from deposits of the cassiterite–sulfide type.

Chalcopyrite. Chalcopyrite forms irregular segregations in quartz and fluorite, generally at grain boundaries. The relations of chalcopyrite with cassiterite and chlorite are unclear, but apparently chalcopyrite is one of the latest minerals in the quartz veins (except for paper spar). The following elements were identified spectrographically in the chalcopyrite: Cu, Fe (n^+); Si, Ca $(0.n^-)$; Pb $(0.0n^-)$; In, Zn, Sn $(0.00n^+)$; Mg $(0.00n^-)$; and Ti (traces).

Chalcosite.   The outer parts of the chalcopyrite segregations are often replaced by chalcosite, and in some cases chalcosite completely replaces chalcopyrite.   In reflected light the chalcosite is grayish white with a distinct bluish cast. It is anisotropic.

Calcite.   Calcite is present as finely lamellar varieties of the paper spar type, often containing very fine inclusions of chalcosite and film segregations of malachite. Often the paper spar is colored brown and reddish brown by iron oxide. The paper spar is the latest mineral in the quartz veins.

It should be noted that, in ores of the third type of mineralization, quartz veins with chalcopyrite, cassiterite, fluorite, and chlorite predominate.

Pyrite Veins. Pyrite veinlets were the last to form. Their thickness does not exceed 5 to 7 mm. They do not have distinct boundaries. They are extremely variable in both dip and strike, and they branch repeatedly and wedge out, grading into linearly arranged grains of pyrite. The following elements were identified spectrographically in the pyrite: Fe (n^+); Si, Al, Mg $(0.n^-)$; Cu, Mn $(0.0n)$; Sn $(0.0n^-)$; W, Zn, V $(0.00n)$; Na $(0.00n^-)$; and Ca and Cr (traces of lines). The pyrite veins generally cut adularia—hematite veins and quartz veins.

Thus, ores of the third type of mineralization display mineral associations characteristic of tin deposits of the cassiterite—sulfide type. The chalcophile aspect of the predominant minor elements in the cassiterite and the wall-rock alteration features of the granite graphically confirm this.

Brief Data on Minerals of the Fourth Stage
of Mineralization

Gangue quartz—fluorite veins of the fourth stage of mineralization cut the entire deposit area in an approximately east-west direction. These veins cut the latticework zones of the first stage or are superposed on these zones and inherit their strike. The quartz—fluorite veins also cut the brecciated ore zone of the second stage and the vein series of the third stage.

The thickness of the veins varies widely, from 0.15 m to 4-4.5 m. Vein boundaries are very sharp. The mineral composition is very simple: quartz, fluorite, finely dispersed hematite, and goethite. Wall-rock alteration of the granite is expressed as weak replacement by quartz and serpentine.

Quartz. The quartz forming the veins is of brownish-gray and violet-gray fine-grained varieties. White comby quartz is invariably present as a discontinuous fringe near the vein contacts. Similar quartz is found in lenticular segregations in the fine-grained metacolloidal quartz groundmass. The violet-gray color of the metacolloidal quartz is caused by finely dispersed hematite.

Fluorite. Numerous crystals of light-violet fluorite showing skeletal forms and also numerous lenticular cavities roughly parallel to the vein contacts are found in the metacolloidal quartz. The walls of the cavities are covered with drusy crusts of clear quartz crystals that often contain inclusions of brownish-yellow acicular goethite crystals forming paniculate and fan-shaped intergrowths. As a rule, radial goethite aggregates are on faces of phantom crystals. Rather large octahedral crystals of colorless fluorite are sometimes noted on the surface of the drusy crust.

Minerals of the Oxidation Zone

There is no oxidation zone as such in the deposit. Supergene minerals present in the deposit are malachite, opal, and finely dispersed iron hydroxides.

Malachite. Malachite is found most commonly in close association with chalcosite; in places it was formed by replacement of the chalcosite. Malachite replacing chalcosite forms dense, finely acicular aggregates that form fringes around chalcosite grains and cut chalcosite segregations with networks of very fine capillary veinlets. Malachite also forms very fine deposits on cleavage surfaces of paper spar. Malachite in the form of sheaflike intergrowths of acicular crystals on drusy quartz in quartz veins of the third stage of mineralization is observed as an exception.

Iron Hydroxides. Finely dispersed iron hydroxides color metacolloidal quartz of the fourth stage of mineralization, and in some cases also paper spar, brownish-yellow and brown.

Opal. Opal, also rather uncommon in the deposits, forming fine capillary veins and films of vein quartz, is more interesting. The thickness of the veins is less than 0.5 mm, and the thickness of the films ranges from hundredths of a millimeter to 0.2 mm.

Rhythmically banded and concentrically zoned structures are the characteristic feature of the opal aggregates. The uneven surfaces of the vein quartz are not always covered by a continuous film of opal; in many cases the opal is segregated in depressions in fracture surfaces of the quartz. The outer parts of the films are rhythmically banded, and the central parts are concentrically zoned. There is a superficial impression of secretion structure, but these structures cannot be explained as layer-by-layer deposition of material, as Pilipenko (1934) explains banding in agates. These structures are caused by internally rhythmic processes: diffusional transfer of pigmenting material (a) from the periphery to the center, to form rhythmically banded structure and (b) from isolated centers of diffusion from the center toward the periphery, to form concentrically zoned structure.

CONCLUSIONS

On the basis of the foregoing, the following conclusions can be drawn.

1. The Shakh-Shagaila deposit is characterized by a multistage ore deposition process. Geologically and geochemically, the first stage of mineralization corresponds to tin ores of the quartz—cassiterite type. The second stage displays transitional features. Paragenetically it is close to ores of the quartz—cassiterite type, but in the structure and morphology of the ores and in the minor elements in the cassiterite it is close to ores of the cassiterite—sulfide type. The third stage corresponds to the chlorite subtype of ores of the cassiterite—sulfide type.

2. Structurally, latticework zones, fractured veins and zones of crushing (brecciated zones) are most characteristic of ores of the deposit.

3. For ores of the first and second stages of mineralization, wall-rock alteration of granite is manifested as greisen formation. Whereas greisenization of the first stage was very intense, the second stage of mineralization shows extremely weak greisenization. Replacement by hematite and chlorite is characteristic of the third stage. Replacement by serpentine and weak replacement by quartz is characteristic of the gangue fourth stage.

4. The mineral associations of each of the stages also vary from those typical of tin deposits of the quartz—cassiterite type in ores of the first stage to those typical of tin deposits of the chlorite subtype of the cassiterite—sulfide type in ores of the third stage.

5. For each type of mineralization the nature of the minor elements in cassiterite varies in full accordance with changes of mineral associations. For cassiterites of the first stage of mineralization the following minor components are characteristic: Fe_2O_3, 0.70%; WO_3, 0.41%; Sc_2O_3, 0.014%; Ti, thousandths of a per cent. In the insoluble residue, Nb (tenths of a per cent) is found. For cassiterites of the second stage the following minor components are characteristic: Fe_2O_3, 0.94%; WO_3, 0.21%; Sb, 0.18%; Cu, 0.023%; Y, 0.02%. The following elements were also identified spectrographically: Ga, $0.0n$; In, $0.00n^+$; Sc, $0.00n^-$. For cassiterites of the third stage the following minor components are characteristic: Fe, n^-; W, $0.n$; In $0.n^-$; Cu, $0.0n^+$; Mn, $0.0n^-$; Ga, $0.00n^+$.

6. The textural and morphological features of the Shakh-Shagaila ores attest to the importance of colloids in the deposition of the ores of the principal stages of mineralization.

Chapter 8

The Iokun'zh Lead–Zinc Deposit

The Iokun'zh deposit, while displaying the basic features of tele-thermal lead–zinc deposits, shows certain special features, mainly (1) good preservation of primary ore textures, (2) absence of any signs of recrystallization and metamorphism of the ores, and (3) morphology of the mineral aggregates (which are primarily pisolitic forms).

The Iokun'zh deposits have been worked for lead since antiquity, judging by the many old workings that are like burrows, the small tunnels, and the shallow pits.

The first detailed information on the deposit was given by Ermakov (1940), who published a geological description of the region itself, with fragmentary data on the individual minerals. The writer (Lebedev, 1954a and b) briefly described the pisolitic aggregates of metacolloidal sphalerite of the Iokun'zh deposit. Arbuzova (1954) has also published some information on the Iokun'zh ores.

GENERAL INFORMATION AND BRIEF GEOLOGICAL DESCRIPTION

The Iokun'zh deposit is in the northwestern foothills of the Darvaz Range, in the upper course of the Obisurkh River, a left tributary of the Iokun'zh River, at an elevation of 2800 m. Administratively, the deposit is in the Khovaling rayon of the Tadzhik SSR, 35 km north-northeast of Muminabad. The main part of the lead–zinc mineralization is restricted to a scarplike ridge of limestone between the Obi-Syrp and Kanioba watercourses.

Upper Cretaceous sediments, including faunally identified Cenomanian, Turonian, and Senonian, are involved in the geology of the deposit region. The Cenomanian consists of alternating red conglomerate and sandstone, clay, limestone, and marl. The Turonian consists mainly of carbonate rocks, of which the thickest are series of sandy limestone and dense oolitic limestone with thin interbeds of sandy clay and marl (shell beds). The Senonian is a thick sequence of interbedded limestone, marl, and clay.

Quaternary deposits, mainly coarse diluvium consisting of angular fragments of conglomerate, sandstone, and limestone, are widespread in the valleys. Igneous rocks are absent from the deposit area.

The Cenomanian, Turonian, and Senonian rocks that contain the deposit form a uniformly dipping sequence that is overturned to the west. They form part of the western limb of the Vasmikukh fan-shaped anticline. Strikes range from northeast (30°) to southwest (120°), and dips range from 20 to 70°.

Faulting, expressed in the deposit area as a series of thrust faults and numerous normal-wrench faults, accompanied the complex folding in the region. The first thrust fault is traced along the base of the main scarp of the ridge, where a thick sequence of Turonian sandstone and limestone is thrust over a sequence of Senonian lime-stone and clay. The second thrust fault is exposed west of the first, between two Senonian sand and clay members. The third thrust is in the scarp of Senonian limestone exposed in the right bank of the Kanioba and is represented by a thick tectonic breccia (up to 25 m) which includes the main lead—zinc mineralization. The rocks of the deposit area are broken into separate blocks by numerous normal-wrench faults that strike north-south and east-west. The dips of the east-west faults are predominantly southwest, and of the north-south faults, west. Displacements range from 1 m to several tens of meters. The normal-wrench fault surfaces are mostly mineralized and form complex networks of calcite veins with sparsely disseminated galenite and sphalerite.

MORPHOLOGICAL TYPES OF MINERALIZATION AND MINERALOGY

Morphologically there are three types of mineralization in the deposit: vein, breccia and disseminated, and lens. These types of mineralization are determined by the nature of the tectonic disturb-ances of the ore-bearing rocks.

Vein Type of Mineralization

Vein mineralization, as mineralized normal-wrench faults with east-west and less commonly north-south strike, is more or less uniformly distributed over the entire deposit area. The veins are extremely varied in mineral composition. Calcite–galenite veins are the most widespread. Calcite veins with galenite and sphalerite are less common, and sphalerite–galenite veins are even less common.

In exposures of gray brecciated limestones in the northeastern part of the main scarp (the zone of the third thrust), there are a great number of calcite–galenite veins, which are associated mainly with normal-wrench faults that strike east-west. The veins are filled with columnar milky calcite with a negligible dissemination of small galenite crystals. The veins are 1 to 5 mm thick. In voids of the thicker veins there are crystals of columnar calcite and galenite. The calcite is represented by ditrigonal prisms terminated by the principal rhombohedron. There are rather large (up to 4 mm) cubic-octahedral galenite crystals, with {100} and {111} equally developed, on the calcite incrustations. The contacts of the veins with the limestone are sharp. No alteration of the limestone near the veins can be observed in the field. Only slight recrystallization of the oolitic limestone is noted under the microscope.

Calcite veins with galenite and sphalerite were found in exposures of brecciated Senonian limestone. These veins differ from the calcite–galenite veins only in having slightly different mineral compositions. A small admixture of sphalerite, as small spherical or less commonly irregular forms, is universally present in the white, coarse-grained calcite mass forming the veins. The size of the sphalerite grains is less than 1 mm. In contrast to galenite, which is in all cases restricted to the zones near the contacts, the spherical aggregates of sphalerite are concentrated mainly in the inner parts of the veins. Occasionally, reniform marcasite segregations are found in the parts of the veins near the contacts and small marcasite globules are found in the calcite and sphalerite. Large spherical marcasite aggregates with globular texture are also found. X-ray study confirms that these aggregates are marcasite.

The calcite–galenite veins observed in exposures in the southwestern part of the ridge of the main scarp, in the saddle, are somewhat different from those described above. There is a large number of old quarrylike workings arranged as benches from northwest to southeast on the northeast slope of the saddle. There are many calcite and calcite–galenite veins with east-west and, less commonly, northwest-southeast and northeast-southwest strike in dolomitized arenaceous Turonian limestone exposed in the quarry walls. The veins are 0.3 to 1.0 cm thick.

Thin zones of fine-grained milky calcite, whose central parts are filled with light-gray translucent calcite with densely disseminated small galenite crystals, extend along the veins near the contacts. By expanding abruptly in thickness, the veins quite commonly form bulges with cavities that are filled with large and well-formed crystals of galenite and calcite. The cavity walls are lined with drusy incrustations of rhombohedral crystals of milky calcite. The thickness of the incrustations ranges from 2 to 5 mm. Galenite crystals represented by combinations of cubes and octahedra, with the octahedra predominantly developed, are unevenly distributed on the incrustations. The size of the crystals varies within wide limits, from 1-2 to 7-8 mm. The surfaces of most of the crystals are covered with a thin brownish-red lead oxide film. Large, transparent calcite crystals represented by combinations of scalenohedra and rhombohedra (the latter truncating the apices of the former) grow on the galenite crystals.

The sphalerite–galenite veins were formed by the filling of fissures with east-west strike, as were the calcite–galenite veins, but they are considerably less common. In exposures of Senonian marly clay and limestone in the northeastern flank of the main scarp are several rather thick sphalerite–galenite veins. Two of these were traced for 80 m down the dip and 30 m along the strike. Their strikes are rigorously east-west, and they are vertical. The thickness of each of the veins, from 2 to 4 cm, is almost constant along both the dip and the strike. The contacts with the wall rock vary; in the gray brecciated limestone they are uneven and wavy, and in the overlying Turonian marly clay and dense yellowish-gray limestone they are very straight.

There are drusy growths and isolated galenite crystals in the parts of the veins near the contacts. The size of the galenite crystals ranges from 1 to 6 mm. The form {100} predominates, with combinations of {100} and {111} common. The drusy growths and galenite crystals are covered by festooned layers of granular sphalerite. The grain size of the sphalerite is not greater than 0.7 to 1.0 mm (the maximum size of the sphalerite crystals in the veins). The sphalerite is light yellow and translucent.

The central parts of the veins are filled with dense, brownish-yellow, finely crystalline sphalerite in which cubic galenite crystals are unevenly distributed. Forms of skeletal growth are observed in individual galenite crystals.

Veins of similar composition were observed in exposures in the southwestern part of the scarp of brecciated Senonian limestones. These veins, though they have the same composition as those described above, show a distinctive beaded structure caused by sharp variations in thickness; at bulges the thickness is 7 to 10 cm, and in narrow places, 0.5 to 1 cm. Sixty lenticular bulges were found in these veins 15 to 16 m down the dip. The distribution of the ore minerals and their interrelationships in these veins are identical to those in the sphalerite–galenite veins described above.

Breccia and Disseminated Type of Mineralization

The breccia and disseminated type of mineralization is found in the southwestern part of the main scarp and in Turonian calcareous and dolomitic sandstone exposed in the southwestern slope of the main scarp. The mineralization is restricted to joints with east-west and northwest-southeast strike accompanied by minor zones of crushing with variable thickness. The wall rock is extremely variable in composition. The proportions of clastic framework and carbonate cement in these rocks vary widely, with clastic material predominant in some places and carbonates in others. Thus, calcareous sandstone grades into arenaceous limestone, and in some cases the latter contains a very small proportion of clastic material. This variation in the composition of the wall rock is repeated over and over both along the strike and down the dip, but it is noteworthy that there is some regularity in the increase in proportions of clastic material in these rocks down the dip (to the east).

The sandy limestone containing the ore is dense, fine-grained, and almost completely dolomitized. Under the microscope it is seen to be a rhombohedral-granular dolomite mass in which rounded and (less commonly) angular grains of quartz and feldspar (microcline and plagioclase) are quite evenly distributed. Small muscovite flakes and rounded glauconite grains are present in minor amounts. The calcareous sandstone, which does not differ qualitatively from this limestone, has high porosity and friability. Its dolomite cement, having completely replaced the calcite cement whose relicts are present in places in the dolomite, is undoubtedly epigenetic. That the proportion of dolomite in the carbonate cement increases sharply to the east concurrently with an increase in the proportion of clastic material and a decrease in intensity of mineralization is also evidence in favor of an epigenetic origin of the dolomite. There are regions in which the carbonate cement consists only of calcite.

The breccia and disseminated type of mineralization is represented by thin galenite–dolomite breccias that are quite variable along the strike, with thickness ranging from 25-30 cm to 2 m. The ore breccias are accompanied by abundant disseminated galenite in the surrounding dolomitized limestone (Fig. 119), which gradually disappears away from the brecciated zone. The width of the aureoles of

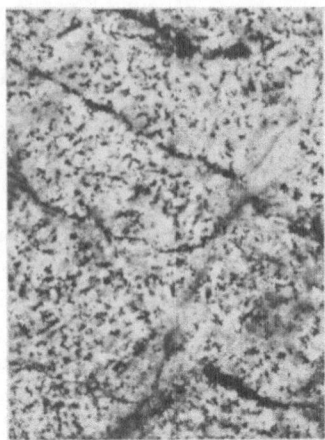

Fig. 119. Disseminated galenite in
dolomitized limestone (× 2).

disseminated galenite and the abundance of the galenite are proportional
to the width of the galenite–dolomite breccias.   Around the thin
breccias (25 to 30 cm) the aureole of disseminated galenite is small,
not exceeding 2 to 3 m; thicker breccias are accompanied by correspondingly thicker galenite aureoles up to 14 to 15 m in diameter.
We shall describe briefly the ore breccias in one of the exploratory
adits.

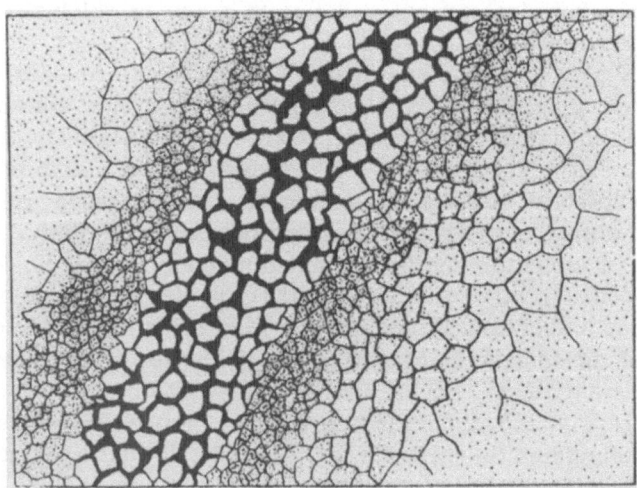

Fig. 120. Structure of galenite–dolomite breccia (shaft no. 12). The
horizontal and vertical scale is 1:10.   Black, galenite; white,
dolomite; stippling, disseminated galenite in dolomite.

The adit was driven across the strike of the ore breccias from the edge of the main scarp into dolomitized arenaceous limestone at the region of most intense mineralization. The adit cuts three breccia zones of east-west strike at 7, 9, and 15 m from the entrance; the thicknesses of the zones are 0.25, 1.20, and 0.30 m, respectively. Figure 120 shows the structure of the breccia ore zones in the west wall of the adit (at 9 m).

A blocky breccia is observed in the periphery of the ore zone. Blocks of dolomitized limestone 25 to 50 cm in size are cemented by galenite that forms a complex network of fine veinlets 1 to 3 mm thick. The limestone here is almost completely altered to dolomite and contains abundant disseminated galenite. Under the microscope the dolomite consists of an even-grained aggregate of rhombohedral grains with rare, rounded quartz grains, relicts of finely crystalline calcite, and xenomorphic grains of disseminated galenite (Fig. 121). The central part of the breccia, quite sharply delimited from the finely crushed material, is characterized by great development of ore cement, which forms 10 to 25% of the breccia. Galenite cements large fragments of dense dolomite that are distinguished from the brecciated dolomite of the peripheral zones by the complete absence of disseminated galenite, distinctive cream-yellow color, and rounded surfaces, caused probably by partial solution of the dolomite by ore-bearing solutions that filled the zone of crushing. Relicts of limestone are wholly absent from this dolomite. A chemical analysis (Table 16) made in the laboratory of the Mineralogical Museum of the Academy of Sciences of the USSR by N. V. Voronkova in 1953 shows that the limestone in the central parts of the ore breccias

Fig. 121. Xenomorphic disseminated galenite (white) in dolomite (dark gray). Reflected light (x 64).

Table 16.   Chemical Composition of Dolomite
of the Brecciated Zones

Component	%	Molecular proportions
SiO$_2$	0.20	—
R$_2$O$_3$	0.32	—
FeO	1.90	00.26
MgO	19.86	0.492
CaO	31.06	0.554
CO$_2$	46.69	1.061
Sum	100.03	—

was completely replaced by dolomite with minor isomorphic iron, which probably governs the color of the dolomite. The following elements were identified in the dolomite spectrographically: Pb and Zn, weak lines; Na and Cd, traces of lines.

Sphalerite is everywhere present in regions with breccia and the disseminated type of mineralization, but it is also present in extremely small quantities as very fine crusts whose thickness does not exceed 1 to 1.5 mm. Morphologically these crusts are dense reniform aggregates and, less commonly, accumulations of well-individualized spherical aggregates. In cross section both the spherulites and the reniform aggregates of sphalerite show radial lamellar structure. The sphalerite is brownish red, sometimes brick red. The following elements are identified spectrographically: Zn, very strong lines; Fe and Mn, strong lines; Cd and Md, moderate lines; Ca, Si, and Al, weak lines; Ge and Fe, very weak lines. All the characteristic lines for sphalerite were present in an X-ray photograph taken in the X-ray structures laboratory of the All-Union Research Institute of Mineral Raw Materials.

In regions where the reniform sphalerite aggregates are not overlain by columnar galenite aggregates, sphalerite is replaced by hydrozincite that forms thin (up to 0.15 mm), silky crusts on the reniform surfaces and the spherulites.

Galenite is present as coarsely crystalline columnar aggregates that encrust rounded fragments of epigenetic dolomite. Well-formed galenite crystals with predominantly octahedral habit are quite common in cavities on the surfaces of the galenite crusts. Combinations of cubes with octahedra and of octahedra with rhombic dodecahedra are less common. Individual crystals are rare; more often they form complex reniform growths. Sometimes parallel growths along the fourfold axis and twins on {111} are observed. The crystals range from 0.2 to 0.8 cm across.

When the large galenite crystals are split along the cleavage, small (0.5 to 0.7 mm) rounded depressions with whitish coatings around the margins are observed on the cleavage surfaces, and freshly split chunks give off quite a strong hydrogen sulfide odor. These depressions in the crystals are arranged in zones and represent fluid inclusions. It was not possible to determine whether one phase or two phases are involved in the inclusions. An aqueous extract was obtained to determine the chemical composition of the liquid inclusions from specially selected regions of the large galenite crystals. Qualitative analysis of the aqueous extract for anions showed: before concentration, the presence of hydrogen sulfide in solution, by reaction with metallic silver, and, after concentration, the presence of chlorine, by reaction with $AgNO_3$ (and in considerable quantity, judging by the great volume of silver chloride precipitate). The following elements were identified spectrographically in the aqueous extract: Pb, Na, and Mg, very strong lines; Ca, Si, and Fe, moderate lines; Mn and Zn, weak lines; Cd and Fe, very weak lines. Thus, primary liquid (or gas and liquid) inclusions whose solutions contain considerable quantities of Mg, Na, H_2S, and Cl are present in the galenite crystals. Table 17 gives a chemical analysis of this galenite made in the ana-

Table 17.   Chemical Composition of Galenite
from the Brecciated Zones

Component	%	Molecular proportions
SiO_2	Trace	—
ΣR_2O_3	0.00	—
MgO	0.02	—
CaO	0.00	—
K_2O	0.00	—
Na_2O	0.03	—
Pb	87.56	4.22
Ag	Trace	—
Zn	0.07	—
As	0.03	—
S	12.93	4.04
Sum	100.65	

Table 18.   Chemical Composition of Cerussite

Component	%	Molecular proportions
SiO_2	None	—
ΣR_2O_3	"	—
MgO	0.94	—
CaO	Trace	—
PbO	83.40	0.371
CO_2	16.35	0.373
Sum	99.79	—

lytical chemistry laboratory of the All-Union Research Institute of Mineral Raw Materials by S. B. Fedorova.

Galenite in regions of breccia and disseminated type of mineralization everywhere shows signs of minor oxidation, but it should be borne in mind that, despite the wide development of oxidation processes in these regions, the intensity with which the hypogene ores were worked through was low.

Crystals and crusts of galenite were covered at the surface by very fine cerussite crusts less than 0.5 mm thick. Crusts up to 1.5 mm thick with well-formed crystals at the surface are observed as an exception.

The cerussite crystals have a tabular habit and display the forms {010}, {110}, and {021}. The crystals are flattened along {010}. Crystals with dipyramidal habit, represented by combinations of rhombic dipyramids {111} and prisms {021}, are less common. The length of the crystals is less than 1 to 1.5 mm along the c axis, and the thickness of the plates is 0.25 to 0.50 mm. Coarse cerussite crystals (up to 1.5 mm along the c axis) represented by combinations of {010}, {110}, and {021} and forming trilling intergrowths on {110} were observed on the surface of the galenite crust in only one instance. The crystals are grayish and semitransparent. Their luster is very strong, approaching adamantine on fractures. The cleavage is distinct along {110} and {021}. The specific gravity, determined pycnometrically, is 6.475.

Chemical analysis of the cerussite (Table 18) made in the chemical laboratory of the Mineralogical Museum of the Academy of Sciences of the USSR by N. V. Voronkova in 1953 showed the presence of Mg and Ca in minor amounts; these elements probably enter isomorphically into the cerussite lattice. The following elements were identified spectrographically in the cerussite: Mn, Zn, Ca, Fe, weak lines; Al, Cu, and Ag, traces of lines.   The heating curve of the cerussite, obtained in the thermal analysis laboratory of the Institute of Geology of Ore Deposits, Petrography, Mineralogy, and Geochemistry, corresponds to the standard heating curve for cerussite.

Lens Type of Mineralization

Lens mineralization is restricted to the tectonic breccia zone of the third thrust fault and is represented by series of small lenticular bodies of sphalerite and galenite–sphalerite composition.

The mineralization is localized in the southwestern flank of the breccia zone.   Thirteen lens-shaped ore bodies of various thickness were identified in a distance of 200 m from the southwestern boundary of the first ledge of the main scarp.   Eight of these are composed of reniform and pisolitic aggregates of galenite and sphalerite, with sphalerite forming about 70% of the ore bodies.   In isolated small lenses galenite predominates slightly over sphalerite, but both in mineral composition and in structure these lenses in no way differ from the others.   Five other ore bodies consist mainly of sphalerite

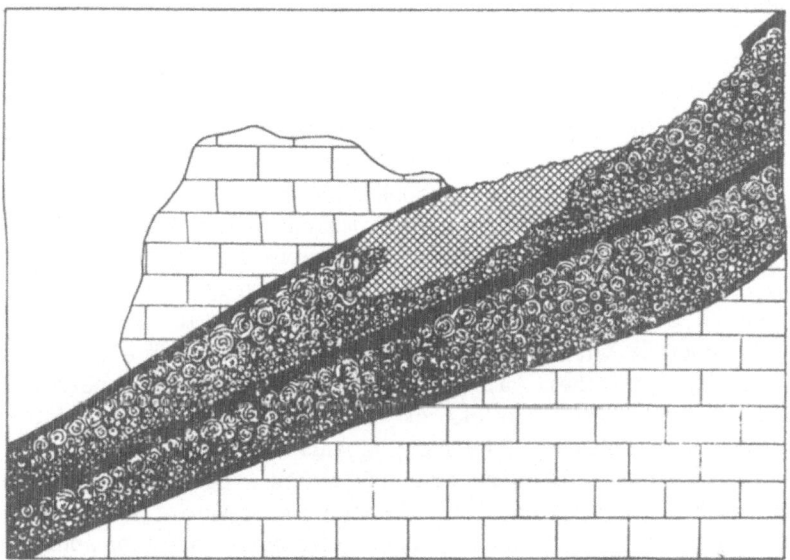

Fig. 122. Structure of sphalerite lens in cross section. Black, dense, festooned sphalerite; main mass, pisolitic sphalerite; cross-hatched region, oxidized ore. Scale 1:100 horizontal, 1:10 vertical.

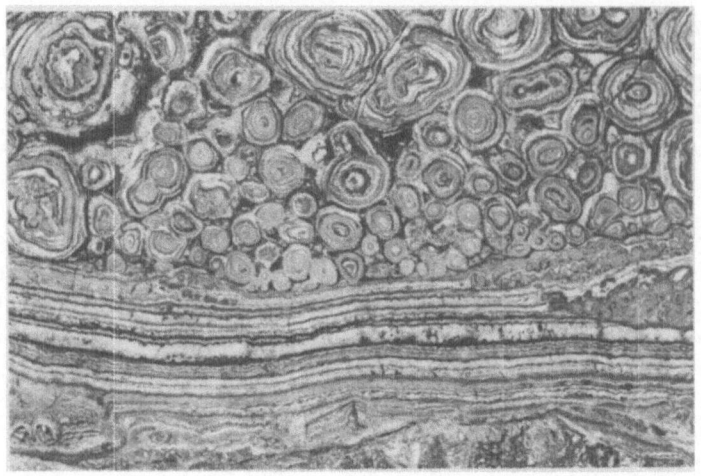

Fig. 123. Details of the structure of a sphalerite lens. Below, layered-
festooned sphalerite of the central part of the vein; above, pisolitic
sphalerite.

in the form of distinctive pisolitic aggregates; galenite is quite
negligible in these aggregates. One of the largest of these sphalerite
lenses is discussed in more detail below.

Ore Lens Composed of Pisolitic Sphalerite

In a natural exposure in the southwestern part of the tectonic
breccia there is an ore body which consists of collomorphic sphalerite
and which forms a strongly condensed, stepwise-curved lens in
dense, greenish-brown, brecciated limestone (Figs. 122 and 123). The
lens dips northwest with the dip of the various segments ranging from
22 to 68°.   The thickness in the central part is 65 to 70 cm; in the
peripheral parts the thickness is no greater than 2 to 3 cm.

The contact with the wall rock is quite sharp; in some parts of
the hanging wall there are joints, expressed as the penetration of
numerous branching veinlets of collomorphic sphalerite into the
surrounding limestone.   The hanging wall is strongly eroded, so that
the lens is exposed over an area of 1500 m^2. Some parts of the exposed surface show signs of solution by ground water. At the surface
there are reniform and concentrically tubercular forms of sphalerite.

Internal Structure of the Lens. In cross section the
lens shows a very distinctive structure. Near the contacts there are
festooned, less commonly hemispherical layers of collomorphic
sphalerite 0.5 to 2 cm thick. Similar layers are also present in the

central part of the lens; these layers differ from those near the contacts by having slightly greater thickness (3 to 4 cm) and by the presence of hemispherical, tubercular surfaces that face in both directions. The two spaces between these layers are filled with sphalerite oolites and pisolites whose size increases upward in each of the spaces (Figs. 122 and 123).

The lens was filled with ore material by two processes; this accounts for the distinctive structure of the lens. First, festooned layers of collomorphic sphalerite were formed in the footwall contact zone, and then pisolites began to form. These pisolites were covered by festooned layers, and then, after a break, this sequence was repeated in the same order. Thus, the lens consists of two "individualized" bodies, upper and lower, and the characteristic structures of the one are repeated in the other.

The pisolites vary widely in size, from 1-3 mm in the lower zone to 3-4 cm in the upper zone. Individual pisolites up to 6.5 cm in diameter were observed. The pisolites are generally quite closely packed. The interspaces are usually empty but are in some cases filled with an aggregate of very small oolites not exceeding 0.3 mm in diameter. Where the spaces between the pisolites are empty, on the surfaces of the pisolites there are thin crusts of tetrahedral sphalerite crystals on which calcite rhombohedra and hydrozincite spherulites are evenly distributed. Much less commonly the spaces between the pisolites are filled with calcite and galenite. The cavities were filled by these minerals after the oolitic aggregates became quite rigid. This is shown by (1) the disintegration of individual oolites and their cementation by calcite and galenite and (2) the truncation of pisolites by galenite veinlets branching from the calcite–galenite void filling. Penetration of galenite between the concentric sphalerite layers in the pisolites is quite common.

Morphology and Texture of the Pisolites. The pisolites forming the lens are predominantly spherical and ellipsoidal. Pisolites with more complex shapes—polyhedral, corrugated, compacted, and irregular (pear-shaped, amoeboid, etc.)—are common. Under more or less tranquil conditions of formation and stabilization of the oolites and pisolites and in the absence of large fragments of wall rock in the cavities where the pisolites formed, pisolites of relatively regular spherical shape developed. For this lens, conditions of relatively tranquil development of pisolites were maintained only in the initial stages of ore deposition. The highly spherical shape of pisolites on the surfaces of the festooned sphalerite layers at the footwall boundary of the lower zone of the lens is evidence of this (Fig. 124).

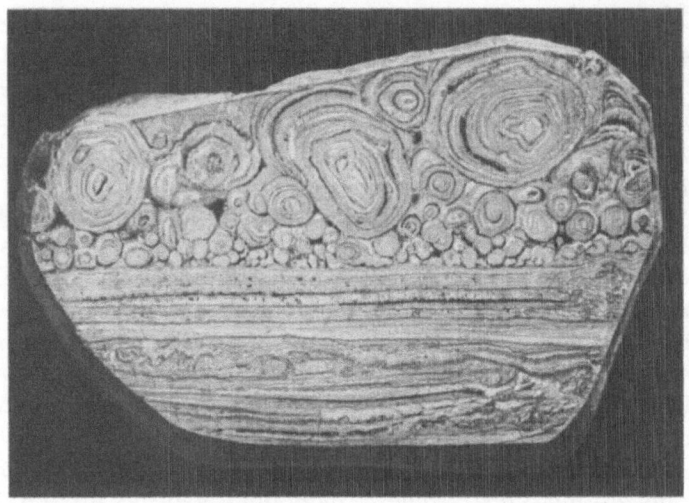

Fig. 124. Pisolitic aggregates and dense bedded forms of sphalerite in the footwall of the lens (x 2).

In addition to the regular spheroidal pisolites of this part of the lens, there are isolated pisolites showing sharp deviations from regular spherical shape—depressed, pear-shaped, etc. The development of these irregular shapes is caused by the envelopment of several small pisolites by an overall cover. In this case, the relations of the sizes of the included pisolites and the orientation of the pisolite aggregates affected the final shape of these complex pisolites. Thus, for example, if the small coalescing pisolites (or oolites) were arranged in a plane, then, when they were surrounded by an overall envelope, pisolites with very flat shapes developed. If the envelope oolites formed an aggregation in which the oolites were oriented variously in space, then the final form of the complex oolite was amoeboid.

The processes of envelopment and increase in complexity of the shape of the pisolites are especially characteristic of the final stages of ore deposition. This is evident in the upper part of the lower zone of the lens, where pisolites with simple shapes are almost absent but various complex shapes are extremely widespread. In these regions, distinctive corrugated pisolites, whose shapes are caused by repeated envelopment of aggregations, are present in addition to the complex pisolites. The corrugated forms are the result of displacement of the pisolites relative to one another, so that the outer shells of the pisolites are deformed and take on a microplicated structure. The shells of the inner zones of these pisolites are usually undeformed.

In the lower part of the upper pisolitic layer regular spheroidal pisolites are quite common, but the main mass consists of pisolites with a great variety of shapes. In the centers of many pisolites forming this layer, there are fragments of broken pisolites and fragments of festooned forms of collomorphic sphalerite. Though these fragments are quite large, the pisolites that formed around them commonly inherited their shape. In this case, polyhedral pisolites of various shapes were formed which in some way reflect the shape of the fragments.

Change in the original shape also took place after the pisolites were formed, during diagenesis. The determining factor in this change in shape is the compaction and settling of the entire aggregate of still viscous pisolites. The result of overall settling of the aggregate is the development of a polyhedral appearance of the pisolites, which at points of contact take on straight boundaries, caused by flattening at individual points on the sphere. Pisolites in the lower parts of each of the two component layers of the lens were deformed most strongly. It is precisely in these parts that pisolites show the most sharply expressed polyhedral appearance, with rounded, rectangular, rhombic, or triangular outlines in section.

In most cases, the surfaces of the pisolites are uneven and hummocky, with microreniform deposits and accumulations of very fine spherulitic sphalerite aggregates; rare, well-formed octahedral galenite crystals and small rhombohedral calcite crystals grow on the sphalerite aggregates. On the surfaces of some pisolites there are clearly expressed radial dehydration cracks dividing the surfaces of the pisolites into irregular dehydration polygons.

In cross section, all of the pisolites show clear concentric structure caused by the alternation of shells of differently colored sphalerite and by the enrichment of some shells in finely crystalline galenite. As a rule, no foreign particles around which deposition could have occurred are observed in the centers of the simple pisolites, and, as noted above, only in the concluding stages of ore deposition did pisolites form by the envelopment of fragments of pisolites that formed earlier and were then crushed. Thus, the centers of growth during development of the original pisolites were probably lumps of globular aggregates which composed the festooned layers underlying the pisolites and which were then still in the gel condition.

There are dehydration cavities of various shapes and sizes in the interiors of many pisolites. There are usually contraction cracks of exfoliation between the individual shells; small cavities of septarian type in the centers of pisolites are also common. Large dehydration cavities of irregular shape are present within the largest pisolites.

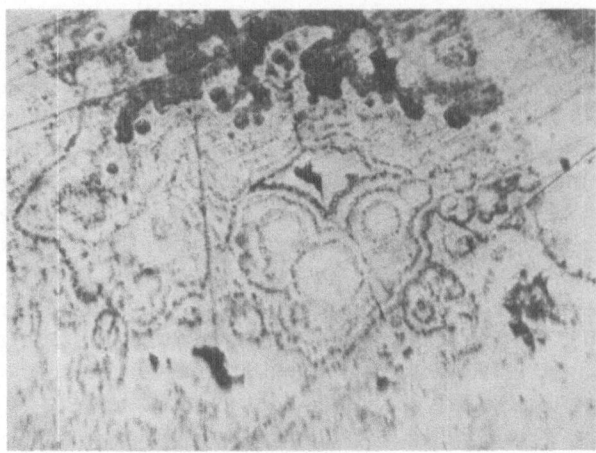

Fig. 125. Microcollomorphic and globular texture of metacolloidal
sphalerite. Etched slightly (× 85).

The walls of these cavities are usually filled with reniform sphalerite
aggregates. The cavities are often partitioned into individual cells
by thin-walled partitions joining the opposite sides of the innermost
shells of the pisolites. The size of the dehydration cavities varies
from 1-2 mm to 3 cm in diameter, depending upon the size of the
pisolites.

Forms Transitional to Reniform Aggregates. In the
upper parts of each of the two divisions, in addition to pisolites with
complex morphology, there are transition zones between these pisolites and the overlying reniform aggregates. Distinctive asymmetric
pisolites, whose asymmetry was caused by oriented deposition of
material to form shells developed on one side, are everywhere present
in these regions. The overlayering of such asymmetric pisolites by

Table 19.   Chemical Composition of Pisolitic
Sphalerite

Component	%	Molecular proportions	Component	%	Molecular proportions
Zn	66.5	1.017	CaO	0.20	—
Cd	0.25	0.002	MgO	0.03	—
Pb	0.55	0.002	FeO	0.20	—
Cu	0.10	0.001	Al$_2$O$_3$	0.00	—
S	32.5	1.013	SiO$_2$	0.00	—
			Sum	100.33	—

festooned layers leads to the development of typical reniform aggregates in the zone near the hanging-wall contact.

Composition and Physical Properties of the Sphalerite. The sphalerite forming the oolitic aggregates is highly dispersed and has an earthy external appearance. It is dirty white to brownish gray. The luster is dull, and the specific gravity, determined pycnometrically, is 4.102. It dissolves easily in HCl with rather vigorous effervescence, which distinguishes it from obviously crystalline varieties and is explained by the high specific surface area of the aggregates.

Under the microscope in transmitted light the sphalerite is not transparent. In reflected light, it is light gray with a rose-violet cast. The reflectivity is low; internal reflections are rose yellow. The mineral is isotropic. It is strongly attacked by aqua regia vapors. After etching, crystalline-granular texture appears very weakly; rare, very small (0.005 to 0.01 mm) crystal grains predominantly of tetrahedral habit are evenly distributed in the mass of microcollomorphic aggregates (Fig. 125).

A chemical analysis (Table 19) made in the analytical chemistry laboratory of the All-Union Research Institute of Mineral Raw Materials by A. E. Popova shows that this sphalerite belongs in the category of iron-poor varieties of cleiophane type, with a typical cadmium content. Galenite in small quantities and chalcopyrite are present as a finely dispersed admixture. The presence of arsenic, thallium, and germanium was established spectrographically.

An X-ray photograph taken in the X-ray structures laboratory of the All-Union Research Institute of Mineral Raw Materials and

Table 20. Interlayer Spacings of Coarsely Crystalline and Metacolloidal Sphalerite

Line no.	Coarsely crystalline sphalerite, Trans-Baikal, Dzhida		Metacolloidal sphalerite, oolitic, Iokun'zh		Line no.	Coarsely crystalline sphalerite, Trans-Baikal Dzhida		Metacolloidal sphalerite, oolitic Iokun'zh	
	d	I	d	I		d	I	d	I
1	3.46	4.5	3.35	5	9	1.482	1	—	...
2	3.13	9	3.05	10	10	1.362	1	—	—
3	2.71	3	—	—	11	1.349	5	1.340	1
4	2.091	5	2.079	4	12	1.238	8	1.233	4
5	1.902	10	1.892	9	13	1.214	1	1.209	—
6	1.788	4	1.783	3	14	1.147	1	—	—
7	1.626	10	1.617	8	15	1.068	10	1.097	6
8	1.554	2	—	—	16	1.020	10	1.036	5

Fig. 126. Flaky and dendritic segregations of galenite in metacolloidal sphalerite (× 64).

computed by G. A. Sidorenko of that laboratory is identical with X-ray photographs of obviously crystalline sphalerites (Table 20).

Types of Galenite Segregations and Their Relations to Metacolloidal Sphalerite. Galenite is a small part of the pisolitic aggregates; it forms not more than 0.5% of the total volume of the ore body composed of pisolitic sphalerite aggregates.

As noted above, the black color of some of the shells in the pisolites and reniform deposits is caused by enrichment in finely crystalline galenite. Even at low magnifications it can be seen that the galenite segregations do not form a continuous aggregate that constitutes all of the shell but instead form thin bands (whose margins have irregular, embayed outlines) that alternate with sphalerite, and also flaky and dendritic segregations in the metacolloidal sphalerite (Fig. 126). It is interesting to note that the galenite segregations are most dense in one particular shell in which up to six or seven of these bands are present and that in the shells nearer the surface there are only flaky or dendritic galenite segregations.

Study under the microscope shows that the banded galenite segregations are distinctive spongy masses consisting of skeletal galenite crystals which form a complex network of interwoven and anastomosing dendritic branches. In many cases there are offshoots of well-expressed skeletal crystals represented by classical shaft forms on the margins of the galenite bands. The concentric arrangement of the banded galenite segregations is probably caused by internal rhythmic processes, namely, diffusional processes in the zinc sulfide gel with subsequent skeletal crystallization of galenite governed by the viscosity of the medium. The flaky galenite segregations at the periphery of the concentric diffusional bands also have spongy texture and show distinctly skeletal growth (Fig. 127).

The writer has observed spongy galenite aggregates with excellently expressed skeletal forms of the component crystals, similar to those described above, in metacolloidal sphalerite samples from the Olkusz group (Poland), Raibl (Italy), and Schmalgraf (Belgium). Apparently the formation of skeletal galenite crystals in the zinc sulfide gel is a general feature of all metacolloidal lead—zinc ores of telethermal type.

Data of Electron Microscopy. Because the sphalerite particles are so highly dispersed, their morphology could be clarified only by electron microscopy. As already noted, in studying the sphalerite under a microscope in reflected light, it was found to be a microcollomorphic aggregate containing small, evenly distributed crystallized regions. The texture of the microcollomorphic and crystallized varieties of this sphalerite and the morphology of the particles forming these aggregates were established by studying the sphalerite by the method of replicas under an electron microscope.

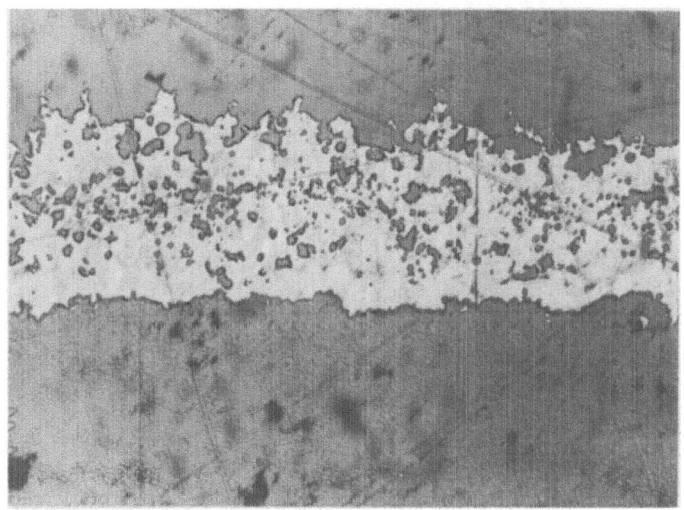

Fig. 127. Skeletal segregations of galenite in metacolloidal sphalerite (× 64).

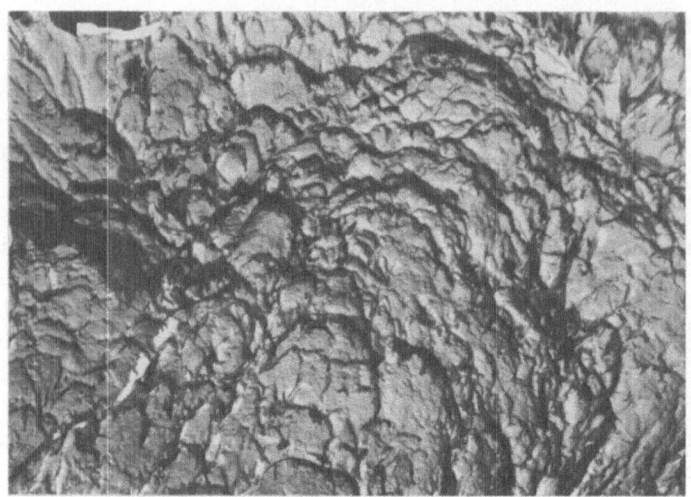

Fig. 128. Structure of the microcollomorphic groundmass of metacolloidal sphalerite. Carbon replica (× 10,000).

The microcollomorphic groundmass of the sphalerite consists
of very small particles 100 mμ to 0.5 μ in size. The particles are
in some cases irregularly rounded and in other cases subhedral and
imbricated.   These two kinds of shapes are related by gradations.
Dense intergrowths of these particles form distinct collomorphic and
reniform aggregates (Fig. 128).

Fig. 129. Structure of crystallized regions of metacolloidal sphalerite.
Carbon replica (× 6000).

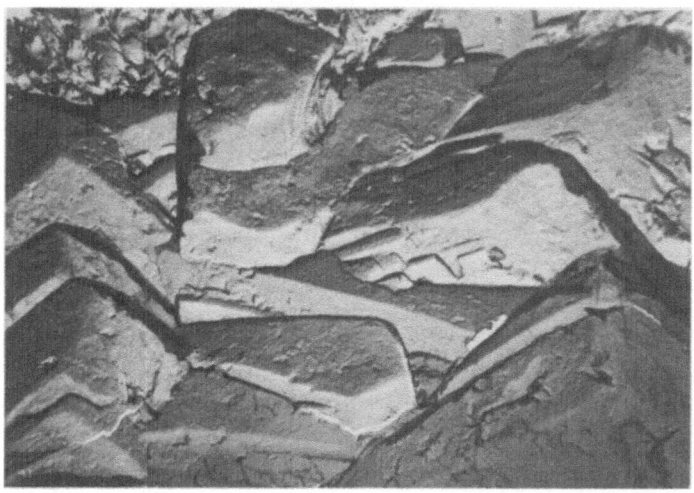

Fig. 130. Nature of the contact between well–crystallized sphalerite and the microcollomorphic sphalerite groundmass. Carbon replica (x 6500).

Regions of crystallized sphalerite are composed of considerably larger crystal particles, 3 to 7 μ in size. The sphalerite particles are well crystallized. The crystals have a tetrahedral or, less commonly, a rhombic dodecahedral habit. Forms represented by combinations of two tetrahedra predominate (Fig. 129). Distinct mosaic structure is often seen on the faces of the crystals (Fig. 130).

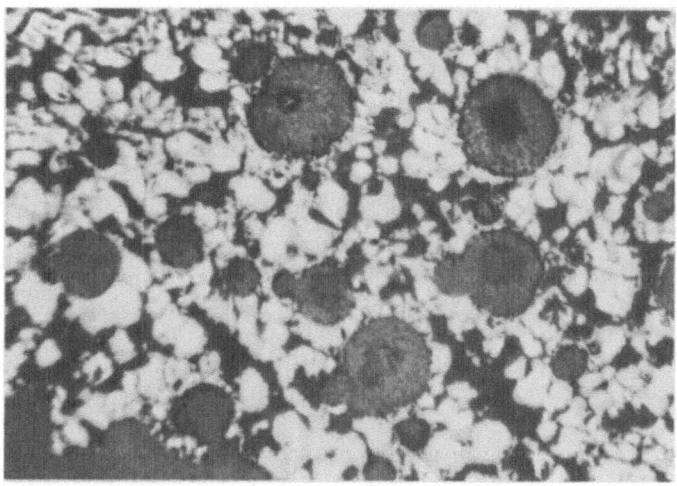

Fig. 131. Spherical forms of metacolloidal sphalerite in dense, fine–grained galenite from the lower bedded–festooned zone. Reflected light (x 800).

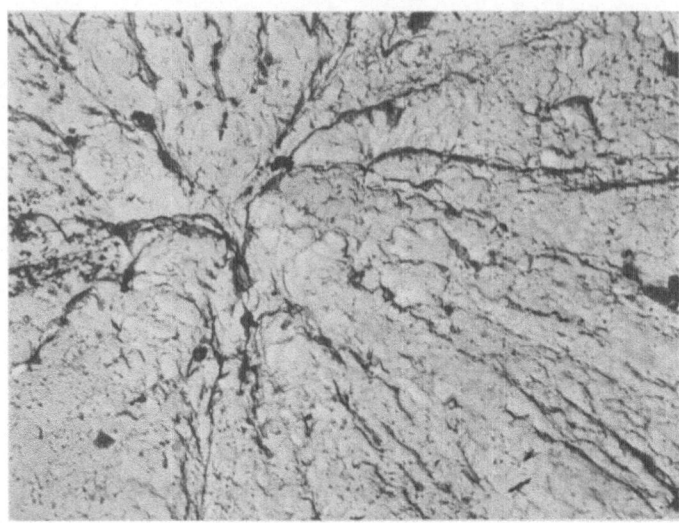

Fig. 132. Structure of metacolloidal sphalerite forming spheroidal forms
in fine-grained galenite. Carbon replica (x 8000).

It is characteristic that, in crystals in contact with finely crystalline collomorphic groundmass, the blocks forming the crystals grade into imbricated aggregates of the groundmass (see Fig. 128).

Thus, the groundmass of the sphalerite consists of aggregates the character of whose particles reflects the initial stages of crystallization of the zinc sulfide gel. Rounded particles characteristic of gels are preserved in the aggregates along with crystallized particles. The aggregates themselves preserve microcollomorphic texture on the whole. Pointlike regions composed of coarser-grained sphalerite evenly distributed in the groundmass are apparently centers of recrystallization.

Microspheroidal forms of sphalerite present in galenite zones in the festooned bands that underlie the pisolites were studied under the electron microscope in addition to the metacolloidal sphalerite forming the pisolites. The textures of these very fine spherical forms of metacolloidal sphalerite segregated in the galenite zones are almost irresolvable with an optical microscope. It can only be established that these forms are aggregates of very small particles that are resolvable with difficulty at the highest magnification (Fig. 131).

Study of these spherical aggregates with an electron microscope allowed several features of their texture to be distinguished. They are composed of very fine particles 0.1 to 0.5 μ in size. The morphology of the particles is distinctive; some are rounded, others have irregular shapes, and still others are weakly crystallized. By intergrowing densely, these particles form granular aggregates with a characteristic texture of mutual boundaries in some regions. On the whole the texture of the aggregates is inequigranular; there is a regular increase in grain size toward the centers of the spherical forms. A euhedral character of the grains is found in these regions. The grain size increases by recrystallization of rounded and irregular particles. By intergrowing, the coarse grains form radially oriented palmate aggregates (Fig. 132).

Radially oriented linear aggregations of very fine particles are also often found. Upon recrystallizing, these linear aggregations form tapered "monocrystals" that retain the radial orientation and are characterized by mosaic structure. The outlines of the elements of the mosaic retain the configuration of the particles forming the chain aggregates (Fig. 133). The particle size increases about a hundredfold in recrystallization. The isolated, largest crystals reach 0.01 to 0.02 mm in length.

Thus, the even-grained metacolloidal sphalerite aggregates that develop by crystallization of zinc sulfide gel are transformed into radial aggregates by recrystallization.

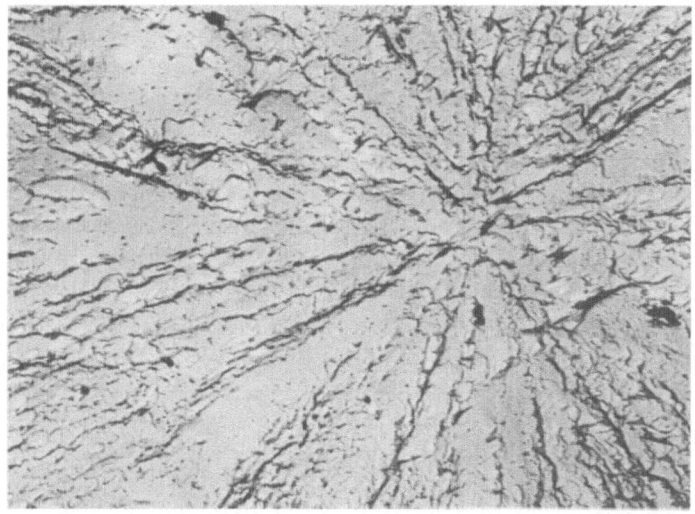

Fig. 133. Radial linear growths of fine particles grading into wedgelike "monocrystals." Carbon replica (× 8000).

Table 21.   Chemical Composition of Partially Oxidized
Pisolitic Sphalerite

Component	%	Atomic and molecular proportions	Component	%	Atomic and molecular proportions
Zn	41.87	0.640	CaO	0.20	0.003
Cd	0.49	0.004	MgO	0.10	0.002
Pb	5.40	0.026	SiO_2	1.12	0.018
Cu	None	—	CO_2	9.80	0.223
ZnO	19.69	0.241	S	22.27	0.694
FeO	None	—			
			Sum	100.94	—

Alteration of Sphalerite in Oxidized Parts of the
Lens.   As noted above, much of the lens is exposed by erosion of
the hanging wall.   On the exposed surface are regions of oxidized
ore, located always along remnants of the hanging wall on the side
from which surface waters flow.   These oxidized regions have the
shape of pockets whose thickness increases toward the remnants
of the hanging wall.

Pisolitic sphalerite aggregates in the oxidized regions, while
preserving their textures, become more friable and lighter in color.
Only the outer zones of the pisolites are altered. The sphalerite is
replaced by conchoidal smithsonite containing more or less evenly
distributed, very small spherulitic hemimorphite aggregates.   Between the conchoidal smithsonite layers there are thin interlayers of
clear yellow greenockite.   Numerous cellular leaching cavities are
filled with friable porous aggregates of powdery sulfur.   In some
places pseudohexagonal cerussite crystals are observed among the
friable pisolitic aggregates.   The sphalerite in the inner zones of the
pisolites is unaltered.

Table 21 gives a chemical analysis of partially oxidized pisolitic
aggregates made in the analytical chemistry laboratory of the All-
Union Research Institute of Mineral Raw Materials by A. E. Popova.
The analysis clearly shows sphalerite, smithsonite, hemimorphite,
and sulfur, thus fully confirming the observed mineral composition.
Comparing this analysis with the data in Table 19 (an analysis of
unaltered sphalerite in the inner zones of the lens), certain regularities of the alteration of sphalerite in the oxidized regions of the lens
become apparent.   In the altered pisolitic aggregates the cadmium
content is increased twofold compared to its original value in the

sphalerite. The lighter color of the altered outer zones of the pisolites is caused by outflux of iron and the light color of the new minerals. Product of supergene alteration of sphalerite are representated by smithsonite and hemimorphite, with smithsonite sharply predominant.

As already noted, pseudomorphic smithsonite, represented by dense, fine-grained, white aggregates with a weak yellowish cast, forms conchoidal aggregates in the outer zones of the altered sphalerite pisolites. In these aggregates there are small relict inclusions of metacolloidal sphalerite and evenly distributed lenticular cells reminiscent of the contraction cracks of exfoliation. In regions of strongest supergene alteration of the pisolitic sphalerite, reniform aggregates of light-gray smithsonite are present in the cavities between friable pisolites. In cross section the reniform aggregates show concentrically conchoidal structure, and individual shells consist of dense columnar aggregates of smithsonite. On fractures the finely columnar smithsonite aggregates have a vitreous to silky luster. One of the cleavage directions appears very clearly perpendicular to the columnar structure. The luster on the cleavage surfaces is weak and pearly. The specific gravity of this smithsonite, determined pycnometrically, is 4.27.

A chemical analysis (Table 22) of the reniform smithsonite made in the chemical laboratory of the Mineralogical Museum of the Academy of Sciences of the USSR by N. V. Voronkova shows the presence of isomorphic Ca and Fe. Hemimorphite inclusions account for the small silica admixture. Weak Mn and Cd lines and traces of Al, Ti, and Cu lines were found spectrographically.

Oxidation products of sphalerite, mainly hydrozincite and much less commonly hemimorphite, are also present in veinlets of collomorphic sphalerite branching off from the main ore body into the

Table 22. Chemical Composition of Smithsonite

Component	%	Molecular proportions	Component	%	Molecular proportions
SiO_2	0.16	0.002	MgO	Trace	—
R_2O_3	0.00	—	FeO	1.42	0.019
ZnO	61.70	0.758	CO_2	35.2	0.800
CaO	1.12	0.027			
	Sum			99.60	—

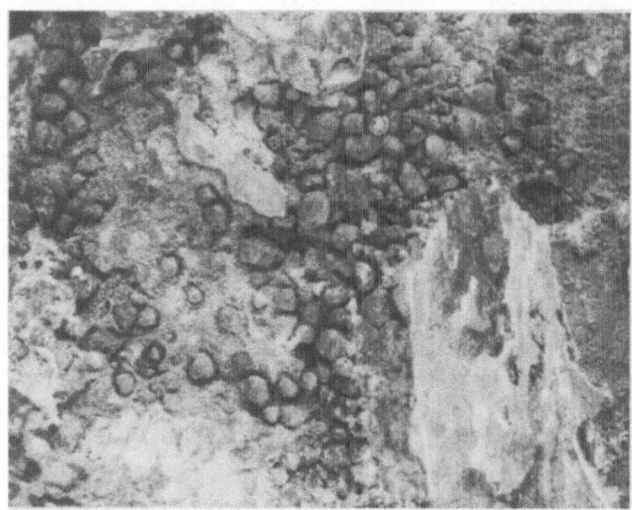

Fig. 134. Separated pisolites of metacolloidal sphalerite cemented
by calcite (x 2).

hanging wall. These minerals are present only in the voids in these
veinlets, and the hydrozincite forms rather coarse, snow-white
reniform aggregates. In cross section these aggregates show radial
and concentrically zoned structure. Under the microscope in transmitted light the hydrozincite is transparent and optically negative.
The indices of refraction are $\alpha = 1.650 \pm 0.002$; $\beta = 1.734$; $\gamma = 1.745$
± 0.002; $\gamma - \alpha = 0.095$.

Thus, smithsonite, hydrozincite, hemimorphite, greenockite, and
native sulfur are formed from the sphalerite in the oxidation zone.
Galenite is not substantially altered.

Structure and Composition of Small Sphalerite Lenses
at the Southwestern End of the Breccia

Several small lenses that also consist of oolitic and pisolitic
sphalerite aggregates, present near the lens described above, deserve
attention. These small lenses are 6 to 8 m below the thick sphalerite
lens in the same very dense and strongly brecciated Senonian limestone. Structurally they differ almost not all from the main lens. A
certain difference is a slightly greater role of calcite in the pisolitic
aggregates forming the lens. Whereas calcite is minor in the main
lens, it forms up to 15% of two small lenses, where it cements separated pisolites of metacolloidal sphalerite. It is slightly less abundant
in two other lenses.

The first lens, extending for about 90 cm along the strike and 15 cm thick in the central part, consists of a dense aggregate of pisolites no different from the pisolitic aggregates described above. As in the main lens, the size of the pisolites increases from the footwall contact toward the center, where they are overlain by parallel wavy layers of dense metacolloidal sphalerite. The size of the pisolites decreases correspondingly from the center to the hanging-wall contact. Parallel wavy layering of metacolloidal sphalerite near the contacts is not present in this lens, and in the lower part of the lens the pisolites are directly adjacent to the dense dolomitized limestone. Spreading of the pisolites and cementation by finely crystalline calcite is found near the hanging-wall contact of the lens.

Two lenses, one 42 cm along the strike and 10 cm thick and the other 35 cm along the strike and 6.5 to 7 cm thick, consist of an aggregate of separated pisolites cemented by calcite (Fig. 134). These lenses are characterized by a single stage of filling and by distinctive structure. In cross section the following upward sequence is observed: An aggregate of separated pisolites cemented by finely crystalline calcite rests on an uneven surface of dolomitized lime-stone. As in the lenses described above, the size of the pisolites increases toward the upper contact. The pisolite aggregate is cemented by milky, finely crystalline calcite occupying about 60% of the lens. In the upper contact zone the pisolites are overlain by parallel fibrous layers of metacolloidal sphalerite; these in turn are

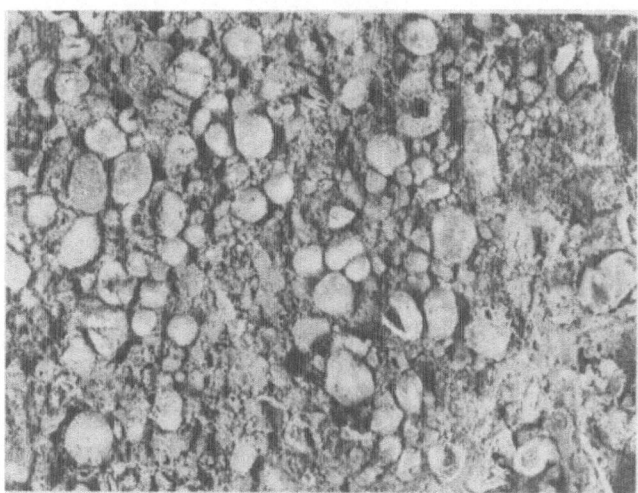

Fig. 135. Hollow pisolites on the surface of dense bedded forms
of metacolloidal sphalerite (x 2).

Table 23. Chemical Composition of Gypsum

Component	%	Component	%
SiO$_2$	0.16	Na$_2$ + K$_2$O	0.31
Fe$_2$O$_3$	0.16	SO$_3$	46.13
CaO	32.99	H$_2$O$\pm$	20.71
MgO	0.13		
		Sum	100.59

The following elements were found spectrographically in the fibrous gypsum: Sr and Zn, moderate lines; Pb and Ba, weak lines; Ga, traces of lines.

overlain by a 0.75- to 1.0-cm layer of coarsely crystalline, semi-transparent calcite.

A fourth lens 40 cm long and about 6 cm thick consists mainly of dense metacolloidal sphalerite. In its central part are small cavities filled with distinctive hollow pisolites (Fig. 135). The spaces between these pisolites are filled with porous, spongy sphalerite. As in the two lenses just described, there is a thin calcite layer at the upper contact.

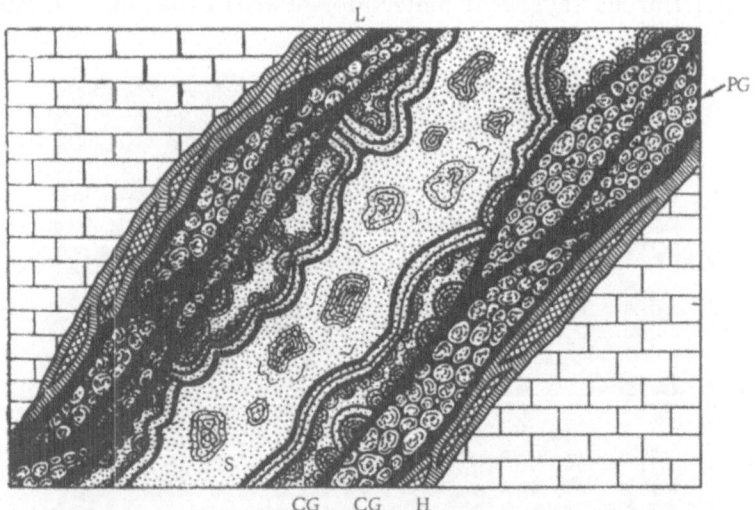

Fig. 136. Structure of galenite–sphalerite lens in cross section. H, hemimorphite; L, limonite; CG, columnar galenite; PG, pisolitic galenite; S, sphalerite. Horizontal and vertical scale, 1:50.

A Lens Consisting of Pisolitic Galenite and Dense
Metacolloidal Sphalerite

A distance of 70 to 75 m northeast of the pisolitic sphalerite lens, in the northeastern drift of one of the exploratory adits (no. 8), there is a small lens of more complex mineralogy and slightly different texture.

The lens is 2.5 to 2.7 m thick in the roof of the drift and 1.7 to 1.8 m thick on the floor of the drift. It dips northwest 28 to 30° and can be traced for 7 m along the dip. Though this galenite–sphalerite lens has textures and structures of its mineral aggregates in common with the sphalerite lens, it is distinguished by certain distinctive textures and different proportions of its component minerals.

The galenite–sphalerite lens is in the same brecciated limestone as the sphalerite lens described above. The contacts with the limestone are sharp. At the contacts there is a lighter zone, 2 to 3 cm thick, in the limestone. Sulfur is present in this lighter zone. Lamellar grains of gypsum that corrode and replace the calcite are seen under the microscope. In places, small lenses of white and light-yellow, columnar, finely fibrous gypsum are present in the limestone being replaced by gypsum. Table 23 gives a chemical analysis of this gypsum (made by N. V. Voronkova).

In parts of the lens near the contacts there is fine-grained yellowish-brown sphalerite that is almost completely replaced by hemimorphite (Fig. 136).

Fig. 137. Dense pisolitic galenite aggregate (natural size).

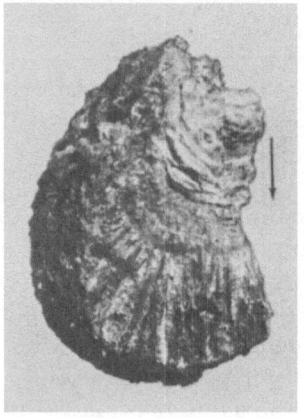

Fig. 138. Asymmetrical galenite–sphalerite pisolite in cross section (arrow shows the direction of gravity; x 2).

Closer to the center of the lens there are zones of pisolitic galenite. Their thickness along the dip ranges from 25-30 cm to 0.5 m. These zones consist of variously oriented, anastomosing crusts of columnar galenite that forms distinctive cells, which are filled with dense pisolitic galenite aggregates (Fig. 136, CG and PG, and Fig. 137). The galenite pisolites are closely packed, which usually disturbs their regular spherical shape slightly. In some places in these regions hollow cells containing individual large

Fig. 139. Structure of galenite zone under microscope. White, galenite; gray, sphalerite (x 64).

spherical pisolites are observed. On horizontal crusts of columnar galenite in the larger cavities are pisolites in the form of distinctive angular hemispheres (see Fig. 22). On nonhorizontal crusts there are pear-shaped and cone-shaped pisolites oriented with the thickened parts downward. These asymmetric forms developed under the influence of gravity by the settling or sagging of sphalerite–galenite pisolites that were in a gel condition.

In cross section the pisolites show concentric zoning caused by alternation of galenite and sphalerite zones. Structural asymmetry, expressed as marked thickening of shells downward, is also observed in cross sections of morphologically asymmetrical pisolites (Fig. 138).

Galenite sharply predominates over sphalerite in the pisolites, and the sphalerite zones are commonly very thin. Equal proportions of galenite and sphalerite are less common. The galenite zones consist of aggregates of radial crystals with characteristic zoning caused by inclusions of metacolloidal sphalerite. The inner sides of the zones show features of skeletal growth of the galenite crystals (Fig. 139). The sphalerite zones consist of dense, yellowish-white, finely dispersed sphalerite similar to the sphalerite filling the central part of the lens.

The outer zones of the pisolites everywhere consist of columnar aggregates of radial galenite crystals elongated along the threefold axis. The spaces between the large galenite crystals in the columnar aggregates are filled with finely crystalline galenite containing rare grains of sphalerite and arsenopyrite.

The surfaces of the galenite pisolites are in most cases rough, less commonly, weakly crystallized. Complex pisolites that are aggregates of several small pisolites (from two to eight) surrounded by an overall envelope are quite common.

Table 24. Chemical Composition of Galenite from the Marginal Zones of Galenite–Sphalerite Pisolites

Component	%	Atomic proportions	Component	%	Atomic proportions
SiO_2	Trace	—	Zn	0.05	0.0007
Fe_2O_3	0.12	—	Pb	86.40	0.412
MgO	None	—	As	0.16	0.002
CaO	"	—	S	13.10	0.409
			Sum	99.83	—

Traces of lines of Cu, Ag, Ca, Mg, and Al were found spectrographically.

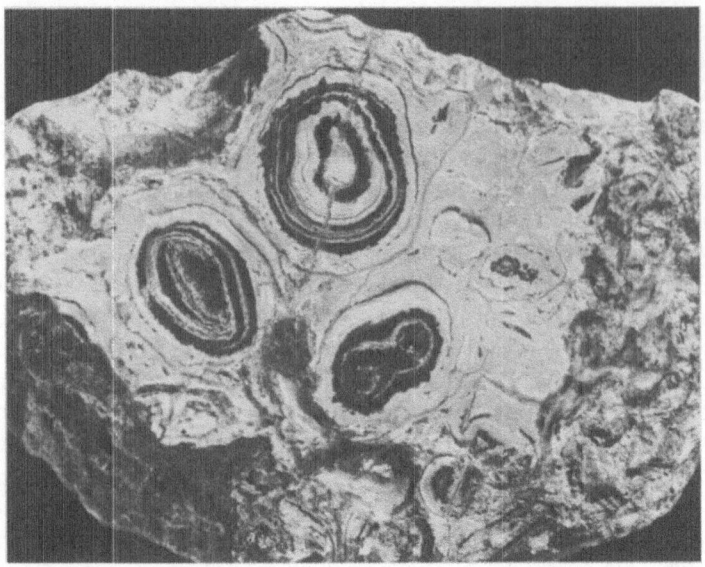

Fig. 140. Concentric forms of galenite in dense, finely dispersed sphalerite
(natural size).

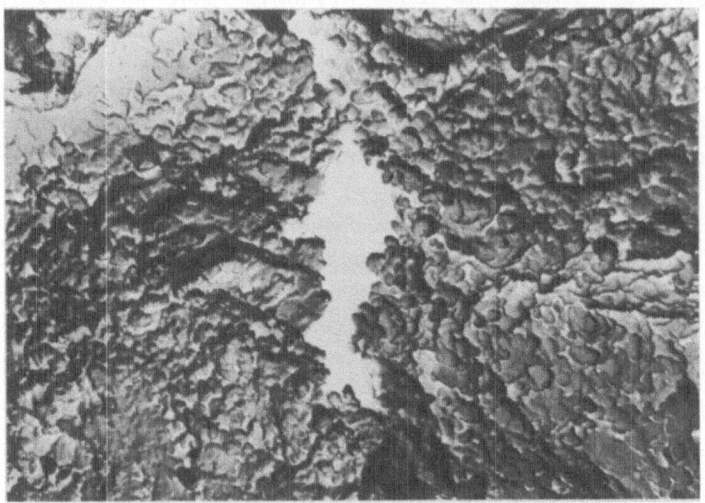

Fig. 141. Globular structure of dense, grayish-yellow sphalerite from the
central part of the galenite lens. Carbon replica (× 12,000).

Table 24 gives the chemical composition of galenite forming the outer zones of the pisolites (analysis made by S. B. Fedorova). This analysis establishes the presence of minor iron, zinc, and arsenic related to the sphalerite and arsenopyrite observed to be present.

The inner crusts of columnar galenite bordering the inner sides of pisolitic zones are covered by reniform aggregates of galenite and sphalerite. These are zoned, owing, as in the pisolites, to alternation of zones of columnar galenite and dense, finely dispersed sphalerite.

The central part of the lens is filled with dense, grayish-yellow, finely dispersed sphalerite containing more or less evenly distributed, fine collomorphic crusts and rare flakes of galenite (Fig. 140). This sphalerite does not form oolitic aggregates, but it is noteworthy that the sphalerite is not uniform. Evenly distributed, dense cores of finely dispersed sphalerite with more friable, partially layered aggregates of sphalerite in the interspaces are manifested very clearly. There is no sharp boundary between the dense aggregates of the cores and the surrounding, more-friable aggregates; the transition from one to the other is gradual.

Only a globular-collomorphic microtexture of sphalerite aggregates is visible under the microscope at highest magnifications. The texture of these aggregates could not be distinguished by optical microscopy, but several textural features could be distinguished by studying this sphalerite under the electron microscope. It was established that the aggregates have predominantly globular texture (Fig. 141). The globule size ranges from 40 mμ to 0.3μ in diameter. In most cases, regions with uniform globular texture are observed in the aggregates. Accumulations of irregularly collomorphic particles formed by the fusion of numerous globules are also common. Regions in which globules form chain and botryoidal aggregations are less common.

Concurrent study of Peruvian brunckite showed that it too, like the sphalerite being described, is a dense globular aggregate. The globule size ranges from 60 mμ to 0.2μ. By combining, the globules in the brunckite also form larger, irregularly collomorphic particles (Fig. 142). Thus, the sphalerite being described is similar to the Peruvian brunckite both in the character of the particles and in the degree of dispersion of these particles.

It should be noted that quite commonly there are signs of crystallization of the globular aggregates in the sphalerite being studied. In the mass of globular and irregularly collomorphic particles there are isolated subhedral particles (Fig. 143). Small uniform regions consisting of imbricated particles similar to those of the pisolitic

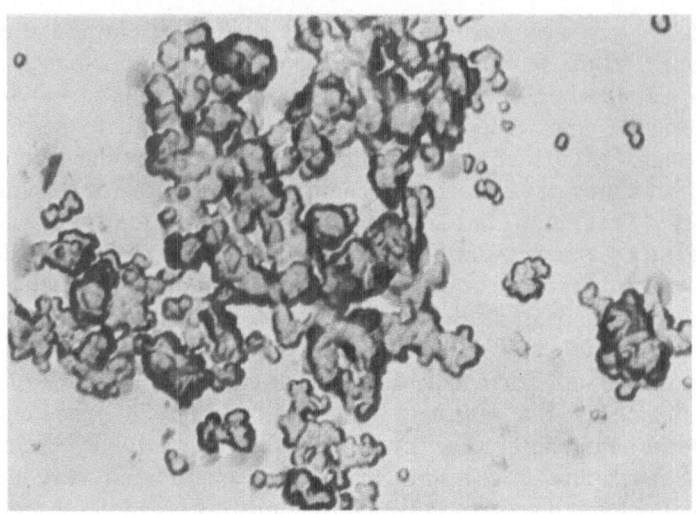

Fig. 142. Globular structure of Peruvian brunckite. Carbon replica (x 24,000).

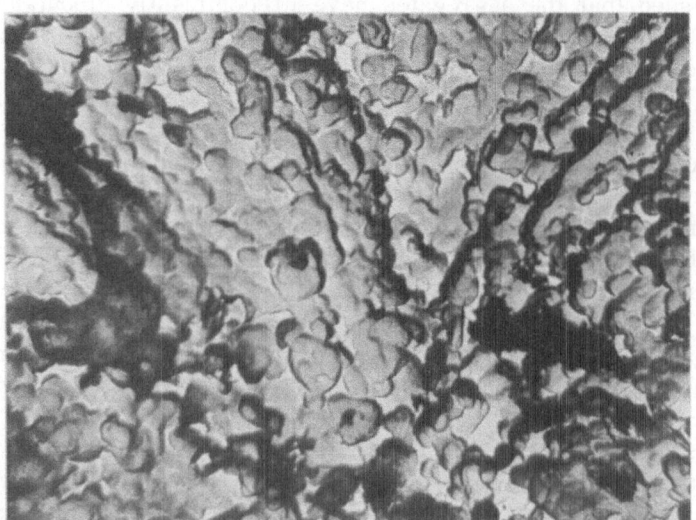

Fig. 143. Globular aggregate of sphalerite with isolated subhedral particles. Carbon replica (x 16,000).

sphalerite are less common (Fig. 144). Individual, clearly crystallized particles are present in these same regions. The most widespread forms are rhombic dodecahedra and tetrahedra. The form
$d\{110\}$ is dominant; combinations of $d\{110\}$ and $o\{111\}$ are sometimes
found, with $o\{111\}$ predominant.

The oxidation products of the sphalerite in the lens are identical to
those of the pisolitic aggregates of the sphalerite lens. The only
difference is that the oxidation is more intense in the galenite—
sphalerite lens.

In the boundary zones of the lens there are crusts and cellular
aggregates of columnar and radial hemimorphite that almost entirely
replace the preexisting fine-grained sphalerite (see Fig. 136, H).
The hemimorphite zones are 4-5 to 30 cm thick. The cavities in the
cellular aggregates are filled with limonite containing small, more
or less evenly distributed lamellar gypsum crystals. The spherulitic
texture of the crusts and cellular aggregates of hemimorphite are
clearly distinguishable under the microscope. Individual large
spherulites of hemimorphite or groups of these spherulites are unevenly distributed in the finely crystalline spherulite mass (Fig. 145).
Relict grains of sphalerite that have irregular outlines and are
strongly corroded by hemimorphite are present in small quantities
in these aggregates (Fig. 146).

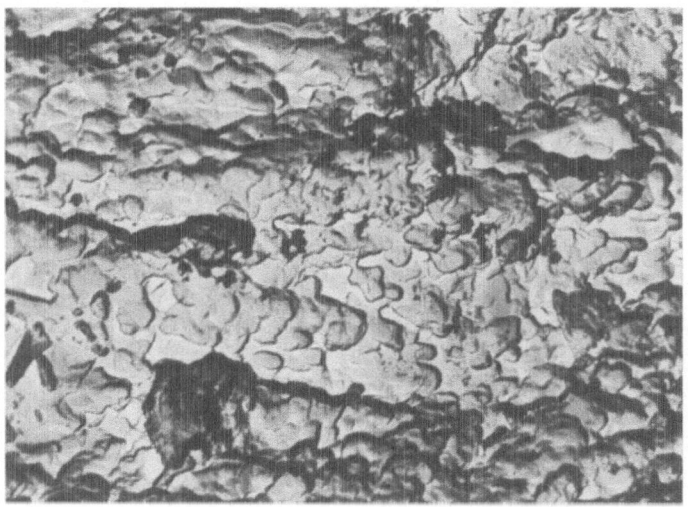

Fig. 144. Uniform aggregate of imbricated particles of sphalerite. Carbon
replica (× 16,000).

Fig. 145. Spherulitic hemimorphite
from near the lens contact (x 40;
crossed nicols).

Small, well-formed tabular crystals of hemimorphite are quite
common in the cavities of the cellular aggregates. Their size does
not exceed 1 mm along the principal axis. The forms {010}, {110},
and {031} are best developed; {100} and {001} are least well developed. The isolated hemimorphite crystals are colorless and transparent; the spherulitic hemimorphite aggregates are white with a
yellowish cast and in places clear yellow. The luster is pearly. There
is good cleavage in two directions, along {110} and {101}. The hard-

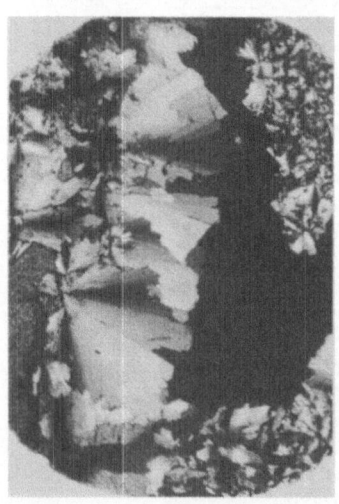

Fig. 146. Sphalerite relicts in a
spherulitic hemimorphite mass
(x 40; crossed nicols).

ness is about 5. The specific gravity, determined pycnometrically, is 3.53. The hemimorphite is decomposed by HCl, with the liberation of gelatinous silica. It is colorless under the microscope in transmitted light. In many grains there are dustlike inclusions of clear yellow greenockite. The mineral is optically positive, and $2V$ determined on a universal stage is $47°$. The indices of refraction are $a = 1.615 \pm 0.001$; $\beta = 1.618$; $\gamma = 1.638 \pm 0.001$; $\gamma - a = 0.023$.

Table 25 gives a chemical analysis of light-yellow hemimorphite made in the analytical chemistry laboratory of the All-Union Research Institute of Mineral Raw Materials by S. B. Fedorova. Sphalerite and greenockite are present as a mechanical admixture in the hemimorphite, and minor Pb and As and traces of Al, Mg, Ca, Fe, and Cu were identified spectrographically. An X-ray picture of this hemimorphite made in the X-ray structures laboratory of the All-Union Research Institute of Mineral Raw Materials is identical with X-ray pictures of hemimorphite from the Gul'shad deposit (Kazakhstan).

Two endothermic reactions, at 475 to 510° and 680 to 690°, and one exothermic reaction at 880 to 930° are clearly shown on the heating curve (Fig. 147) obtained in the laboratory of thermal analysis of the Institute of Geology of Ore Deposits, Petrography, Mineralogy, and Geochemistry. The first endothermic reaction is related to the liberation of water of crystallization, with a weight loss of about 4.3%; the second is related to the liberation of hydroxyl water, with a weight loss of 4.4%. The total weight loss is 8.7%, which corresponds quite closely to the percentage of volatiles in the analyzed hemimorphite, 8.53%. The strong exothermic effect is probably caused by crystallization of willemite.

Strong oxidation of sphalerite is also observed in isolated regions of the lens consisting of galenite–sphalerite pisolites. The sphalerite,

Table 25. Chemical Composition of Hemimorphite

Component	%	Molecular proportions
ZnO	67.20	0.826
SiO$_2$	24.78	0.412
Cd	0.043	0.0003
S	0.25	0.0077
H$_2$O$^\pm$	8.28	0.460
Sum	100.553	—

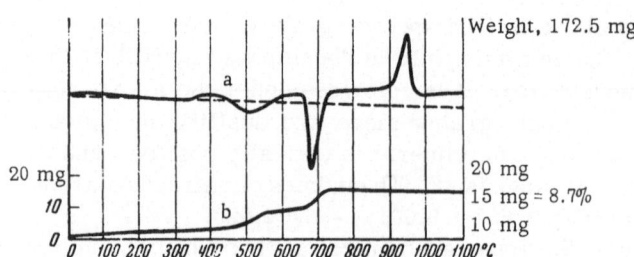

Fig. 147. (a) Heating and (b) weight-change curve of hemimorphite.

which generally forms the centers of the pisolites, is often replaced completely by friable aggregates of very small hemimorphite spherulites.   Individual cerussite crystals, represented by dipyramidal forms of pseudohexagonal habit, are present in these aggregates. There are very fine veins of hemimorphite and smithsonite in the sphalerite forming the central part of the lens. Where the network of hemimorphite veinlets is most dense, the sphalerite is strongly altered and replaced by a fibrous hydrozincite aggregate. Thus, both the process of oxidation of sphalerite in the lens and the oxidation products formed are identical with those of the pisolitic aggregates of the sphalerite lens.   The difference lies in the slightly higher intensity of these processes in the present case.

The foregoing data show that in the Iokun'zh deposit there was a tendency toward spatial segregation of the two main components of the ores, sphalerite and galenite. The main mass of the sphalerite was deposited in the first stage of ore deposition and was localized within the thick tectonic breccia of the third thrust fault. Sphalerite is the predominant mineral in the lenticular ore bodies within this thrust zone, and despite the moderate abundance of galenite in some lenses the zone of tectonic breccia of the third thrust fault can be characterized as a zone of substantially zinc ores. The metacolloidal character of the ores of this zone, which are products of crystallization of zinc sulfide gel and, in part, gels of mixed composition with sharply varying proportions of ZnS and PbS, deserves emphasis. The main mass of the galenite was deposited in the concluding stages of ore deposition in zones of crushing in the normal-wrench fault scarp of Turonian arenaceous limestone and calcareous sandstone.   The galenite is of well-crystallized varieties; it forms disseminated grains, a network of monomineralic veinlets, and crusts with excellently expressed drusy surfaces. Sphalerite is negligible in the zones of crushing, and the zone of breccia and disseminated mineralization can be characterized as a zone of predominantly lead ores.

BRIEF CONCLUSIONS

The lead—zinc mineralization of the Iokun'zh deposit is restricted to thrust faults and normal-wrench faults related to Alpine orogeny. The correlation between intensity of faulting and intensity of mineralization is clearly manifested in the deposit area. The strongest faulting was in the zone of the third thrust fault and in the southwestern part of the deposit, in Turonian limestone, and it is precisely to these regions that the principal mineralization of the deposit is restricted.

No direct connection between mineralization intrusion has been established, but it is noteworthy that 7 km southeast of the deposit in a Jurassic-Cretaceous evaporite sequence there is a dike of porphyritic andesite containing regions of intense serpentinization with disseminated galenite and chalcopyrite. There is no doubt that the lead—zinc mineralization of the Iokun'zh deposit is genetically related to Alpine magmatism.

It does not yet seem possible to solve the problem of the modes of transport of ore components, but it is quite likely that the lead and zinc sulfides were carried in complex ionic-colloidal solutions in which the dispersing media were true solutions of alkali-metal (mainly sodium) and magnesium sulfides and the dispersed phase consisted of finely dispersed particles of ZnS and PbS and possibly also double sulfides. Probably separation from the melt and the original transfer took place in the form of an aerosol (Chukhrov, 1950), which later graded into a hydrosol of various concentration. Ore deposition from these solutions was by coagulation of the dispersed phase, caused by the appearance of an active coagulating agent, in the form of calcium ions, in the solutions.

Of the number of features that distinguish the Iokun'zh ores from metacolloidal ores of other deposits of telethermal type, the following are noteworthy:

1. The predominance of oolitic and pisolitic aggregates of sphalerite and, less commonly, galenite and sphalerite.
2. The finely crystalline nature of the metacolloidal sphalerite, which is almost identical with brunckite (Herzenberg, 1938, and Sterk, 1953).
3. The good preservation of globules and spherical aggregates of globules (globulites) that form the main mass of the weakly crystallized metacolloidal sphalerite aggregates.

Chapter 9

Experimental Data on Diagenetic Changes in Zinc Sulfide Gels and Two-Component (ZnS + PbS) Gels

The wide distribution of metacolloidal sphalerite in nature is well known. It is most widespread in young lead–zinc deposits, where it generally forms dense massive varieties with characteristic collomorphic texture. These are usually recrystallized, but in most cases relicts of oolitic and globular texture are preserved in the crystallized aggregates. Deposits of almost unmetamorphosed, finely dispersed sphalerite forming aggregates of very fine globules (Cercapuquio, Cerro de Pasco, Peru), or aggregates of oolites (Truskovets) or pisolites (Iokun'zh) are also known.

On the whole, the patterns of formation of globular and oolitic sulfide ores remain unclear. In view of this, the writer made experiments on the diagenetic alteration, with time, of zinc sulfide gels and two-component (ZnS + PbS) gels and the production of spherical aggregates in the laboratory. During the experimental work such matters as the effect of electrolytes on crystallization of the gel and the conditions of formation of polymorphic modifications of the zinc sulfides were also studied.

DIAGENETIC CHANGES IN ZINC SULFIDE GELS WITH TIME

To study diagenetic changes, zinc sulfide gels were produced and then aged by stages. The experiments were made in open containers by precipitating zinc sulfide from aqueous zinc sulfate solutions at various concentrations by adding highly concentrated sodium sulfide solutions. The high concentration of the solutions of the precipitating agent necessitated knowing the minimum oxidizing effect of atmospheric

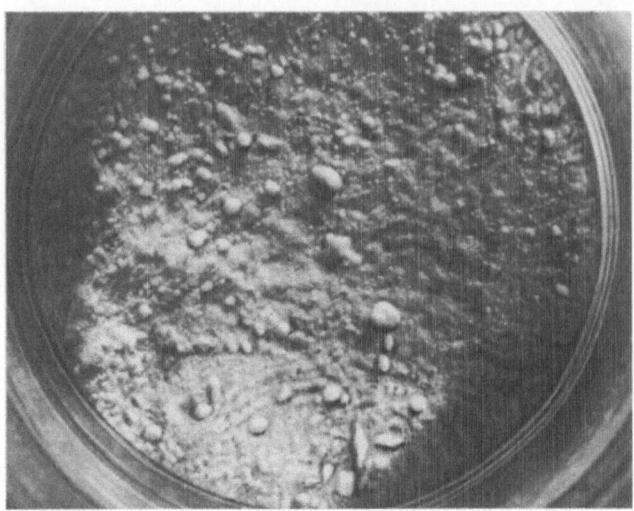

Fig. 148. ZnS oolites on the surface of a globular gel of the same
composition (in solution) (natural size).

oxygen, if only at the time of precipitation and in the first stages of
diagenesis of the precipitate.

Precipitation was usually done in three or four batches. In the
first precipitations, small portions of the precipitating agent were
added (for example, 15-20 ml to 500 ml of a 3% $ZnSO_4$ solution), and
in later precipitations an excess was added. In adding the first
portion of precipitating agent to the 3% $ZnSO_4$ solution, a bulky precipitate was immediately formed at the surface of the solution.
After 10 to 15 minutes the precipitate decreased sharply in volume
and slowly sank to the bottom of the container. There were also
changes in the surface morphology of the amorphous precipitate; for
the first few minutes the precipitate was friable and did not have a
distinct surface of separation from the solution, but upon compaction
the boundary became more clearly expressed by taking on a smooth,
rounded, or sometimes reniform shape.

Study of the precipitate under a binocular microscope showed that
it consisted of very small, rounded concretions, 0.1 to 0.5 mm in size,
more or less evenly distributed in a uniform gelatinous mass. The
shape of these concretions was usually strictly spherical, less commonly ellipsoidal. Upon addition of subsequent portions of precipitating
agent, the size of the spherical concretions increased markedly to
3 to 4 mm in diameter.

Heating of both the initial solutions and the precipitates to 80° produced no change in the texture of the precipitate. However, heating of the precipitate to 100 to 110° and brief boiling of the solution with the precipitate destroyed the globular texture and transformed the amorphous precipitate into a crystalline precipitate.

The amorphous precipitates were held in the mother liquor for two months. Then most of the solution was carefully decanted, and the remaining part was diluted with distilled water to the previous volume and evaporated at room temperature (18 to 20°). Evaporation was continued for one to one and one-half months, depending on the volume of the solution above the precipitate. During this time the evaporating solution was twice cleaned of the sulfur film that formed by oxidation of hydrogen sulfide. Generally a film no longer formed after the second cleaning (toward the end of evaporation).

After the solution was evaporated, the container, with a still very moist precipitate, was tightly closed and the precipitate was aged in humid air for a month. Then the container was opened and the sediment was aged in dry air. During the first month of this subsequent aging the precipitate was repeatedly washed with distilled water to remove soluble salts, mainly sodium sulfide, and finally was dried and held in the same medium for one additional month.

Thus, the sediment passed through three stages of diagenesis: in the mother liquor, in humid air, and in dry air. The final product was a thick, yellowish-gray, gluelike aggregate forming more than 80% of the total volume of the precipitate. Its upper part was more friable and graded into a very friable white aggregate of fine oolites (Fig. 148), which always occupied the upper 15 to 20% of the precipitate.

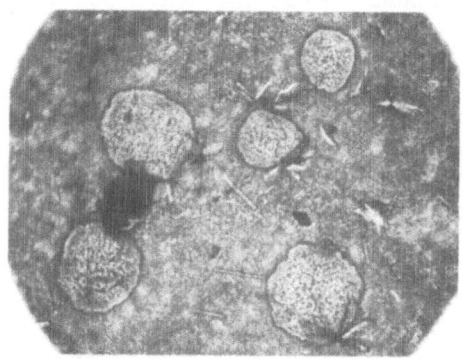

Fig. 149. Sphalerite globulites in a mass of finely globular gel (x 20; uncrossed nicols).

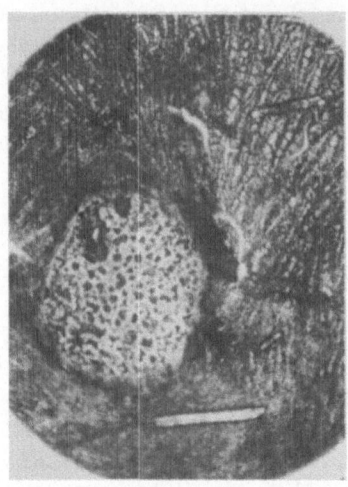

Fig. 150. Sphalerite globulite in
a mass of chain aggregates of ZnS
globules (× 50; uncrossed nicols).

Study of specially prepared thin films of partially dehydrated zinc
sulfide gel under the microscope showed that the thick, gluelike
material was a uniform, extremely fine-grained mass which at high
magnifications showed globular texture.   It is noteworthy that quite
often the closely packed globules formed chain aggregates grouped in
parallel or fan-shaped bundles (Fig. 150).   The cause of these chain
globular aggregates remains unclear.   Larger spherical forms
(globulites) were observed in this finely globular aggregate: rounded
concretions of very fine globules surrounded by a thin, very dense gel
envelope (Figs. 149 and 150).   The globulites were more or less
uniformly distributed in the finely globular gel, and they gradually
became more abundant toward the outer parts of the precipitate.

Table 26. Chemical Composition
of Zinc Sulfide Gel

Component	%	Molecular proportions
Zn	53.76	822
S	25.61	800
H_2O^-	20.30	—
Sum	99.37	—

The surface layer of the precipitate consisted of friable, coarsely globular aggregates containing large (up to 4 to 5 mm) sphalerite oolites.

After aging six months in dry air a dense, finely globular aggregate with rare inclusions of large globulites was chemically analyzed in the analytical chemistry laboratory of the All-Union Research Institute of Mineral Raw Materials by A. E. Popova (Table 26). The analysis showed that the precipitate was zinc sulfide and contained abundant adsorbed water.

X-ray work on the dense, finely globular aggregate (Table 27, a), done after six months of aging, showed that the precipitate was completely amorphous. An X-ray powder photograph of a friable aggregate consisting of globulites and oolites that was aged six months shows four wide and diffuse sphalerite lines (Table 27, c). After another six months of aging (i.e., after a full year of aging) these same aggregates were again X-rayed; a powder photograph of the dense, finely globular aggregate shows five wide, distinct sphalerite lines (Table 27, b), and a powder photograph of the friable globular oolitic aggregate is substantially similar to that of metacolloidal sphalerite from Iokun'zh (Table 27, d and e).

Table 27. Interlayer Spacings of ZnS Gel After Six to Twelve Months' Aging*

Line no.	a d	a I	b d	b I	c d	c I	d d	d I	e d	e I
1			3.473	5	—	—	3.479	5	3.35	5
2			2.174	10	3.029	10	3.081	10	3.050	10
3			2.125	3	2.070	5	2.113	5	2.079	4
4			1.920	7	1.890	10	1.909	10	1.892	9
5			—	—	—	—	1.822	4	1.783	3
6			1.630	6	1.616	3	1.613	9	1.617	8
7			—	—	—	—	1.346	4	1.340	1
8			—	—			1.233	6	1.233	4
9			—	—	—	—	1.157	1	1.209	1
10			—	—	—	—	1.101	7	1.097	10
11			—	—	—	—	1.040	2	1.036	5
12			—	—	—	—	0.950	3	—	—
13			—	—	—	—	0.906	5	—	—

(Column a: Amorphous to X-rays)

*Notation: a, dense colloidal ZnS with finely globular structure after six months' aging; b, the same after one year; c, friable, globular-oolitic ZnS aggregate after six months' aging; d, the same after one year; e, metacolloidal sphalerite of the Iokun'zh deposit.

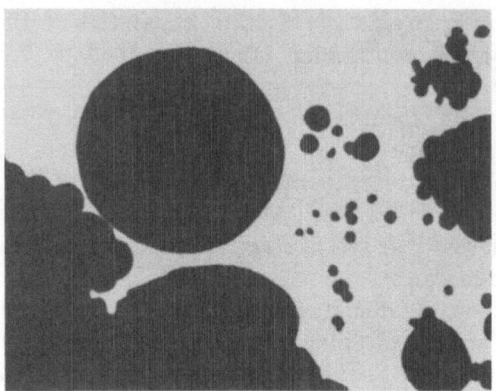

Fig. 151. Globular structure of ZnS gel aged six
months. Suspension method (x 18,000).

Study of the zinc sulfide gel under an electron microscope showed
that the dense gels that were aged six months were uniform globular
aggregates with predominantly fine (0.1 to 0.3 μ) globules, among which
considerably larger globules were unevenly distributed. In many
cases the smaller globules were clumped around the larger globules
(Fig. 151). Partially coalescent globules, both small and large, were
observed. It should be noted that such "hardened," only partially

Fig. 152. Aggregate of coalescent globules (ZnS gel aged six months).
Carbon replica (x 16,000).

Fig. 153. Initial stages of crystallization of globular aggregates of ZnS. Suspension method (× 18,000).

coalescent globules were already present in viscous gels that were aged one and one-half to two months in the solutions (Fig. 152). This is evidence that the process of the coalescence of globules had died out by the end of the first stage of diagnesis.

Also, there were a considerable number of globulites, uniformly distributed in a homogeneous globular mass, in dense zinc sulfide gels that were aged for six months. Large globules in the initial stage of

Fig. 154. Crystalline particles of ZnS formed by crystallization of a globular gel (see Figs. 151 and 152). Carbon replica (× 11,000).

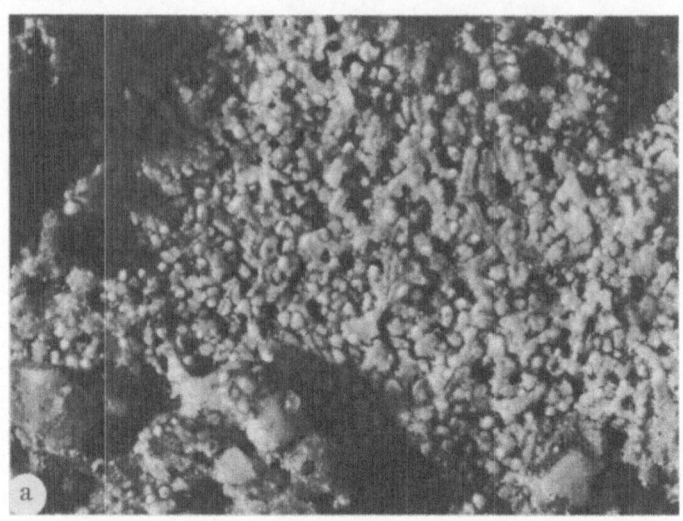

Fig. 155. Oolitic aggregates of metacolloidal cassiterite . (a) Obtained
experimentally (× 25); (b) natural (Iokun'zh deposit) (× 20).

crystallization were sometines observed in these aggregates. In this case, their spherical shape was not disrupted.

A small, acute-angled protuberance that later grew into a dendritic crystalline skeleton was formed at the surfaces of the globules (Fig. 153).

Thus, the ZnS gel retained its colloidal nature for the first six months. Diagenesis was expressed as coalescence of globules (the earlier stage of diagenesis) and formation of globulites, more complex spherical aggregates (the later stage of diagenesis). Isolated centers of crystallization developed. The aggregate as a whole was still amorphous with respect to X-rays, and only the regions with globular-oolitic texture near the surface showed wide and diffuse sphalerite lines in the X-ray photographs.

After one year of aging the particles of these same gels differed sharply from those described above by showing well-defined crystal outlines. The particle boundaries were not always straight but were sometimes denticulated and stepped; this was probably caused by skeletal growth of sphalerite crystals. Sharply terminated sphalerite crystals represented by combinations of two tetrahedra, {111} and {1$\bar{1}$1}, were often observed (Fig. 154). There were distinctive particles with dendritic outlines between more or less well-crystallized large particles.

Thus, crystallization of the zinc sulfide gel into sphalerite develops quite clearly with time. In the conditions of the given experiment, crystallization did not destroy the collomorphic textures acquired by the precipitate in the gel condition.

EFFECT OF ELECTROLYTES ON RATE OF CRYSTALLIZATION OF THE GEL

Considering the importance of calcium chloride and sodium chloride in the composition of the gas and liquid inclusions in endo-genic sphalerites, both well-crystallized and collomorphic, experiments to clarify the effect of these compounds on the structure and rate of crystallization of the gels were made. In precipitating the zinc sulfide, a small quantity (from 0.1 to 0.4%) of Mg and Ca chlorides were introduced into the zinc chloride solution. The effect of high calcium chloride concentrations on the diagenesis of the zinc sulfide gel was also studied.

It was found that adding even small quantities of concentrated NaCl and $CaCl_2$ solutions to the zinc sulfide sol caused the sol to

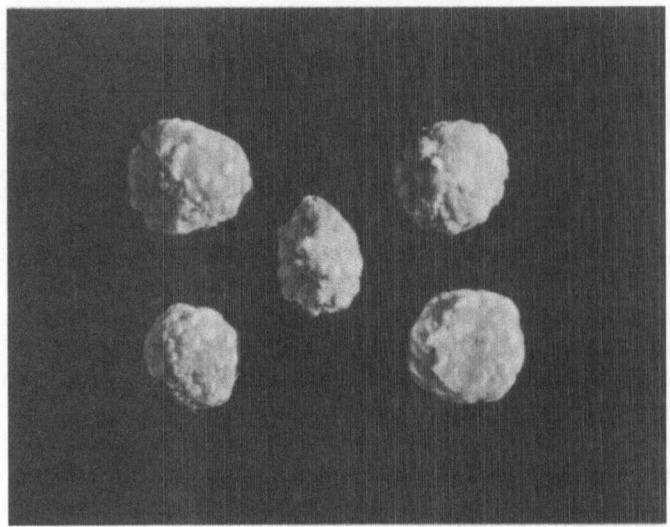

Fig. 156. Large sphalerite oolites obtained in saturated chloride solutions (x 5).

coagulate rapidly; also, adding small quantities of NaCl and CaCl₂ to the original solutions from which the zinc sulfide was precipitated caused rapid globulation of the gel. When NaCl and CaCl₂, particularly the latter, were present in the original solutions, immediately after precipitation the gel took on a distinct granular (microglobular) texture.   Repeated precipitation of zinc sulfide from these solutions usually led to the formation of considerably larger oolites than those obtained from solutions without these compounds (Fig. 155). The mean size of the sphalerite oolites obtained under these conditions ranged up to 4 to 5 mm, but isolated oolites were 7 to 8 mm in diameter (Fig. 156).

In aging the precipitate in the mother liquors containing NaCl and CaCl₂ (0.25%), after one and one-half to two months the precipitates became solid, and the individual oolites became firm and stable under changes of conditions of the ambient medium, especially under sharp variations of air temperature and humidity.   We note that it was precisely under these conditions that oolites with well-expressed concentric zoning were obtained; the outer shells of the oolites were distinctly crystallized into a columnar aggregate (Fig. 157).

Powder photographs of oolitic aggregates obtained in the same way and aged for one and one-half months in the mother liquor show all the principal sphalerite lines.

Because the time required for experiments on diagenesis of the gels was long (four to five years) it was possible to determine the effect of electrolytes on recrystallization. It should be emphasized that recrystallization is very important in the diagenesis of the gels, especially in the later stages.

Observations made over five years on "dry"* zinc sulfide gels, both with NaCl and $CaCl_2$ (0.25%) and without, showed that both change gradually into crystalline aggregates. However, the course of the recrystallization was different in the two cases.

In gels without NaCl and $CaCl_2$, ZnS crystals 0.02 to 0.03 mm across were observed only at the third year. Certain features of re-crystallization of the gel precipitate are not without interest. At the end of the second year of aging, evenly distributed darker spots about 1 mm in diameter appeared in the overall uniform mass of the pre-cipitate. The number of these spots was one or two per 3 cm^2. Over the next six months the size of these spots increased to 5 to 6 mm, and isolated spots grew to 8 mm. Study of the precipitate from these regions in immersion liquids showed under the microscope that, in con-trast to the gel surrounding the globular mass, the zinc sulfide gel in these regions consisted of very fine tetrahedral crystals hundredths of millimeters in size. Thus, the spot regions were formed by incipient recrystallization of the gel in isolated centers more or less uniformly distributed in the precipitate. A year after the appearance of the numerous recrystallization centers, the size of the recrystallized regions increased to 1.5 cm and many neighboring regions "blended" with one another. The crystals grew to 0.2 to 0.3 mm. It should be noted that recrystallization never proceeded to conclusion in this experiment. The recrystallized regions formed not more than 55 to 60% of the total volume of the gel precipitate.

The recrystallized gel precipitates in whose precipitation sodium and calcium chloride, in quantities not exceeding 0.25%, were introduced into the solution differ substantially from the recrystallized "pure" gel. Numerous point centers of recrystallization appeared in them as early as eight months after they were obtained, i.e., after six months' aging of the precipitate in dry air. They were more than ten times as abundant as those in the "pure" gels and formed ten to twelve centers per square centimeter of observed surface. In this case the size increase of the recrystallization centers was determined every ten days, and, on the twentieth day after they appeared, 70% of the centers had increased sharply in size and had merged with one another.

*It should be noted that, even three years after the precipitates were obtained and after two and a half years of aging in dry air, the ZnS precipitates contained 14 to 18% adsorbed water.

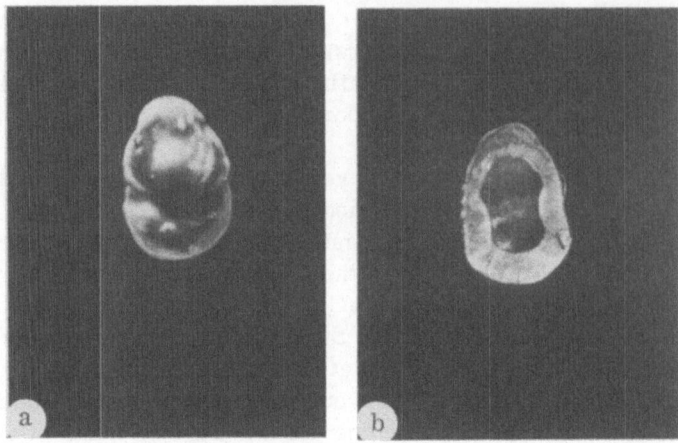

Fig. 157. Large sphalerite oolite with finely columnar outer zone
(× 10). (a) Surface view; (b) cross section.

Accordingly, about 70% of the total volume of the precipitate was recrystallized.   Control observation of the precipitate under the microscope confirmed the presence of crystalline particles 0.05 to 0.1 mm
in size.   After another 15 days the entire precipitate was recrystallized into a finely granular aggregate.

Thus, during nine to ten months of aging in dry air, the zinc sulfide
gel containing a small admixture of electrolytes had recrystallized
completely.   Also, periodic checking under the microscope in the
following years showed that even after two years the grains of the
aggregate were still growing, so that recrystallization was in fact
continuing.   At the end of the third year of aging, the size of the
crystalline grains in the aggregate had reached 0.8 to 1 mm. After
this the process began to die out, and toward the end of the fourth
year there was no noticeable increase in the grain size.

Rare, solitary sphalerite grains 1.5 to 1.8 mm across were found
only in isolated regions of the aggregate.   The sphalerite grains in
the aggregate were asymmetric and perfectly transparent.   Many
were partially terminated; one or two tetrahedral faces were present
in addition to mutually grown boundaries.   Well-formed crystals,
usually combinations of two tetrahedra, {111} and {11$\bar{1}$}, were less
common.   Twinning on {111} was sometimes noted under the microscope in polished sections; polysynthetic twinning was not observed.

Comparing the course of recrystallization* of the dry gel pre-

*Aging of the gel precipitate in a dry air medium of one sort or another was carried out
with other conditions the same. The volumes of the initial precipitates were the same;
the temperature was 18 to 20°; the adsorbed water content was 17 to 20% for the pure gels
and 14 to 18% for gels containing electrolytes.

cipitates of zinc sulfide with and without a small admixture of electrolyte, the rapidity and intensity of recrystallization in the first case and the weakness and incompleteness of recrystallization in the second case is quite apparent. The foregoing data show that NaCl and $CaCl_2$ admixtures definitely have a strong effect on the rate and intensity of crystallization and subsequent recrystallization of the zinc sulfide gel.

The features of diagnesis in a ZnS gel with a high concentration of coprecipitated calcium sulfide* by obtaining mixed ZnS + CaS gels with proportions of ZnS : CaS = 5:2 were studied. These gels were obtained by precipitation with the aid of sodium sulfide from mixed 3% solutions of $ZnCl_2$ and $CaCl_2$. The precipitate very rapidly (four to five days) became separated into two layers: a light-gray lower layer of thicker consistency and an upper, white, more friable layer.

After one month of aging in the mother liquor the layered gel decreased sharply in volume, but the upper layer remained considerably less dense than the lower layer.

Study under an electron microscope showed that crystalline particles predominated in the upper layer. Globular forms constituted not more than 10 to 15% of the gel. Among the crystalline particles were large, elongated crystals and, rarely, extremely small cubic crystals. The larger particles are prismatic gypsum crystals 10 to 20 μ long and 1 to 3 μ across (Fig. 158). The forms b {010}, l {111}, and m{110} were determined visually. The crystals were elongated along m{110}. The small crystalline particles had distinct cubic form and apparently corresponded to residual NaCl that remained in small quantities in the preparation after washing. The cubic particles, 0.1 to 0.4 μ in size, were sharply subordinate to gypsum.

Rare globules of sphalerite formed pear-shaped aggregates 0.15 to 0.3 μ in size; as a rule, individual isolated globules did not exceed 10 mμ in diameter.

After another month of aging (in dry air) the upper layer of the precipitate was transformed into a solid crust composed of very small gypsum crystals. Under the electron microscope it was seen that the size of these crystals was sharply greater, the smallest of them being 40 to 60 μ long and 7 to 10 μ across. Gypsum crystals often included sphalerite globules during growth (Fig. 159). The greater part of the sphalerite globules formed interstitial fillings between large gypsum crystals (Fig. 160).

The lower, denser layer of the gel precipitate consisted mainly of zinc sulfide. Under the electron microscope the precipitate was

---

*In this experiment it remained unclear whether calcium sulfide was precipitated together with zinc sulfide or whether a mixture of calcium polysulfides was formed which by rapid oxidation transformed into gypsum.

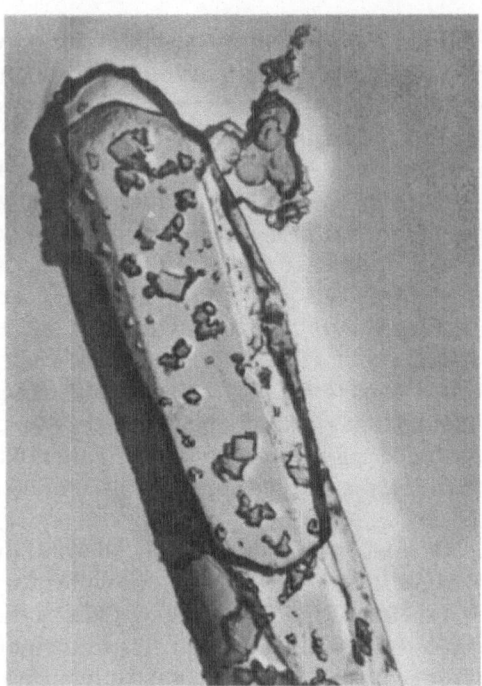

Fig. 158. Gypsum crystal with globules and globular aggregates of ZnS clinging to its faces (the small cubic crystals are residual NaCl). Carbon replica (× 9000).

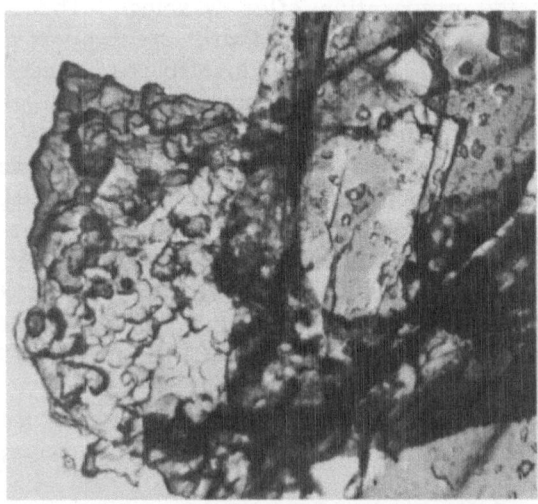

Fig. 159. Capture of ZnS globules by gypsum crystals during growth of the crystals. Carbon replica (× 9000).

seen to have uniform globular -texture, similar to that of the zinc sulfide described above.

It is noteworthy that numerous contraction cracks and exfoliation cavities were formed upon sharp decrease in the volume of the precipitate during the one month of aging in the mother liquor. At first, thin cracks parallel to the surface of the precipitate were formed in the overall mass of the gel. Then over two to three days they expanded quite rapidly and took on a lenticular outline. In this, the inner surfaces of the cavities, which at the beginning were flat and even, gradually became reniform. At about the twentieth day of aging, these forms, while still retaining a viscous gel consistency, developed into distinct reniform aggregates (Fig. 161). Upon further compaction of the gel, certain changes could be noted on the surfaces of the large (3 to 4 mm) hemispheroidal elements. These elements themselves shrank slightly, and their smooth, even surfaces, corresponding to almost perfect hemispheres, became bumpy and ultimately divided into segmented bulges. After two months of aging, the reniform aggregates became very dense and solid, as did the entire mass of the gel.

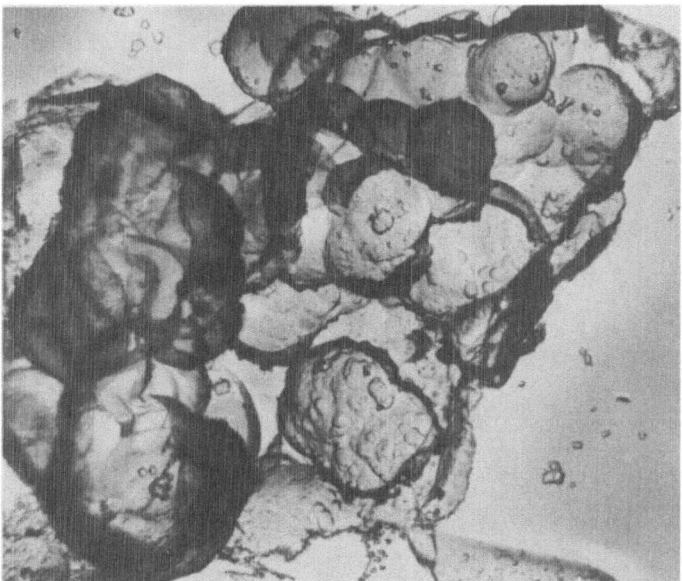

Fig. 160. Globular-reniform aggregate of metacolloidal sphalerite forming interstitial filling between gypsum crystals. Below is the surface of a gypsum crystal; above and to the right are imprints of gypsum crystals. Carbon replica (× 9000).

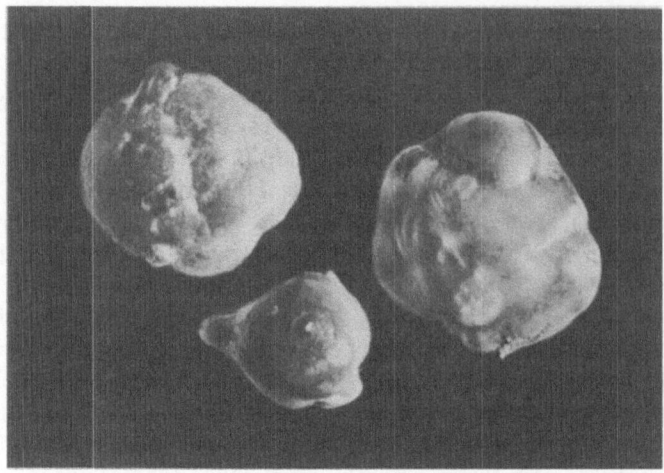

Fig. 161. Separated reniform elements of ZnS forming the surface of
contraction cracks in ZnS gel (× 12).

Fig. 162. Structure of ZnS aggregate on a radial fracture surface of a
reniform element. Carbon replica (× 21,000).

A carbon replica of a radial fracture surface of the hemispheroidal elements was obtained in order to reveal the structure of the ZnS aggregate forming them. Electron-microscope study of the preparation showed the complete absence of any crystallized particles in the aggregate. The aggregate consisted of densely packed, very small (0.05μ) globules and spheroidal aggregates of globules (i.e., globulites) 0.2 to 0.4 mμ in diameter (Fig. 162).

The layered gel precipitate of zinc sulfide and gypsum did not change substantially over the next four years of aging in dry air. Only the gypsum recrystallized. This recrystallization was quite intense; by the end of the fourth year the gypsum crystals were 1 to 1.5 mm long and 0.1 to 0.2 mm across on the average. No signs of recrystallization were noted in the zinc sulfide. Electron-microscope data showed that the texture of the sphalerite was globular as before, and only in isolated regions did coalescent globules form microreniform aggregates (Fig. 163). Crystallized particles were absent.

Thus, high calcium chloride concentrations in the initial solutions and calcium sulfide precipitated along with the zinc sulfide did not accelerate crystallization of the ZnS gel. On the contrary, the reverse

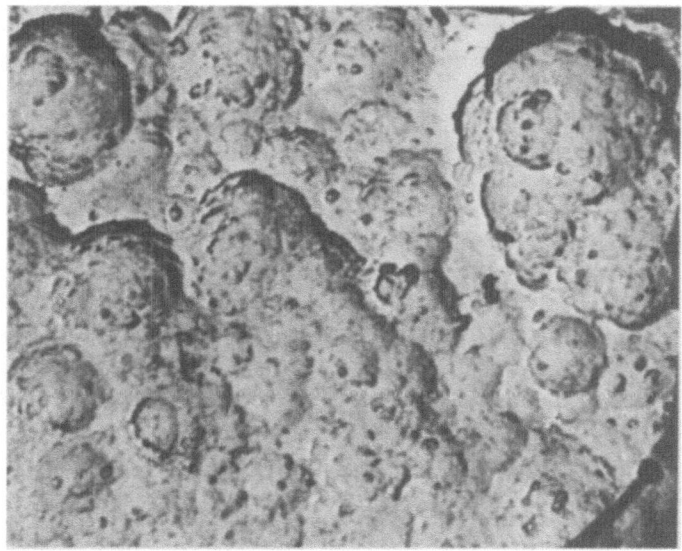

Fig. 163. Globular-reniform aggregate of ZnS from interstitial fillings between gypsum crystals after four years of aging. Carbon replica (× 25,000).

was the case; the initial texture that formed in the earlier stage of diagenesis of the gel tended to be preserved.

In the given experiment, these compounds not only had a strong stabilizing effect on the texture of the aggregate but also promoted energetic condensation of the gel precipitates. In this, contraction cracks and exfoliation cavities whose surfaces rather rapidly became reniform were formed more often than under simple conditions.

The foregoing experimental data show that the effect of electrolytes on the rate of crystallization and subsequent recrystallization of the zinc sulfide gel varies depending on the concentration. Low concentrations promote crystallization and intensify recrystallization. High concentrations usually suppress crystallization and recrystallization.

DIAGENETIC CHANGES OF TWO–COMPONENT (ZnS + PbS) GELS

In order to study the aging of mixed ZnS + PbS gels, gel precipitates with variable proportions of these components were produced. The aging conditions of these two-component gels were similar to those of the zinc sulfide gels. The two-component gels were obtained by alternately adding 3% solutions of zinc chloride and lead acetate to a highly concentrated sodium sulfide solution. A uniform gray precipitate was thereby formed. Separation into layers, with the segregation of monomineralic regions, started even at the very beginning of the first stage of aging. Separation into layers proceeded differently depending upon the proportions of components.

Aging of Mixed ZnS + PbS Gels with ZnS Predominant

We shall first discuss certain patterns of aging of ZnS + PbS gels when zinc sulfide predominates. A mixed gel with the proportions ZnS : PbS = 10:1 was obtained by the above method. After 10 hours this outwardly homogenous gel was studied under an electron microscope by the suspension method. Examination of the preparation showed that the mixed two-component gel had a uniform globular texture. The particles were very small globules, both separate and as chain aggregates. There were no particles with crystal outlines.

During the first five days after precipitation, the gel became thick and the first signs of layering appeared. The entire mass of initially homogeneous gray gel separated into three differently colored layers:

dark gray, lower; light gray, middle; and yellowish white, upper. Differentiation of the precipitate reached a maximum after another 10 days; the lower layer became black, the middle layer became spotted (dendritic segregations of lead sulfide became segregated in the white zinc sulfide gel), and the upper layer remained yellowish white. Thus, over 15 days the initially homogeneous two-component gel became sharply differentiated into three layers: sphalerite (5 cm thick), sphalerite–galenite (3.5 cm thick), and galenite (2 cm thick).

The layered gel was held in the mother liquor for four months, after which, samples from each layer were taken. By this time the gel precipitate had contracted sharply in volume, and the thickness of all three layers had decreased almost twofold. After sampling, most of the mother liquor was carefully decanted, and 150 ml of distilled water was added to the small remaining part. The distilled solution was again decanted. This was done three times, and then a small part (25 to 30 ml) of the remaining solution was evaporated at room temperature. Subsequent stages of aging of the differentiated two-component gel were similar to those described for the zinc sulfide gel.

By studying the samples under an electron microscope it was possible to recognize the nature of the particles in the various parts of the layered sphalerite–galenite gel after four months of aging in the mother liquor.

The sphalerite gel forming the upper zone of the layered precipitate consisted of numerous chain globular aggregates and incompletely coalescent clumps of two, three, or more globules. Individual, isolated globules, ranging in size from 0.1 to 0.5 μ (Fig. 164a), were less common. The partially coalescent globules formed particles up to 1.5 tp 2 μ in size.

As in the upper zone, the sphalerite and galenite particles in the sphalerite–galenite gel of the middle zone were in the form of globules. Isolated large (1 to 1.5 μ) globules were common, and chain aggregates of globules were less common. Galenite was present as particles with crystal outlines (Fig. 164b). Skeletal forms, all the way to skeletal dendrites, predominated. The character of the particle outlines is evidence that the crystalline skeletons formed by strong rib growth of cubic-octahedral forms. The size of the galenite particles was from 1 to 4 μ.

The galenite gel forming the lower zone of the precipitate consisted of relatively large crystallized particles (Fig. 164c). Their outlines corresponded to octahedral and cubic-octahedral forms. Elements of skeletal growth of the particles appeared as regularly oriented fine serration of the outlines. Clearly expressed skeletal forms were

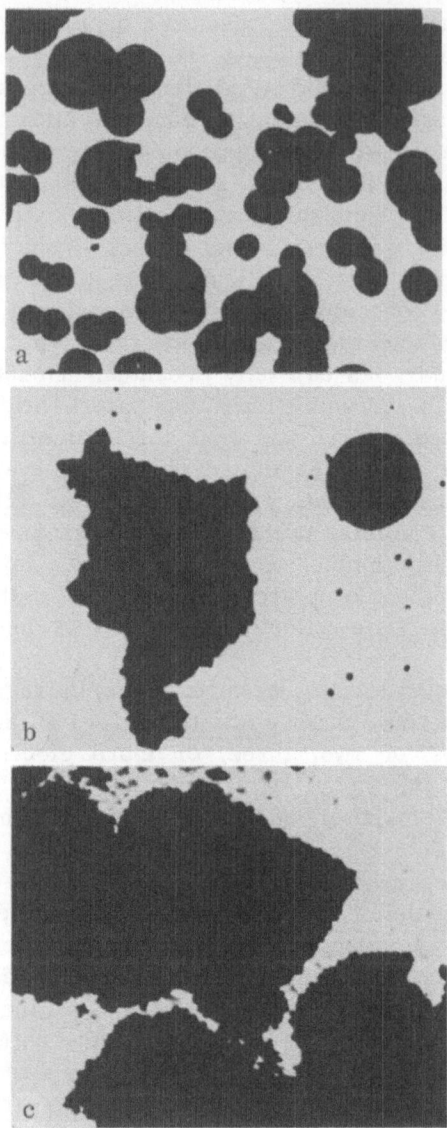

Fig. 164. Nature of particles in layered sphalerite–galenite   gel.   Suspension   method
(x 10,000). (a) Particles of the upper zone of
the precipitate; (b) particles of the intermediate
zone; (c) particles of the lower zone.

rare. The size of the galenite particles in this zone ranged from 2 to 7 μ. Particles 4 to 4.5 μ in size predominated. There were sometimes aggregates of coalescent particles 10 to 15 μ in diameter. It is noteworthy that sphalerite particles, in the form of individual large globules, were extremely rare in this zone.

Thus, even at the stage of early diagenesis, the patterns of later alteration of the precipitate were already in evidence.

The mixed ZnS + PbS gel globulated immediately after precipitation. The particles of both components were exclusively globules. Over 24 hours the mixed gel retained both its external uniformity of color and its internal uniformity of texture.

The subsequent separation of the gel into layers was caused by crystallization of the lead sulfide globules and growth of the particles thus formed. The increasing size of the galenite particles accelerated separation of the mixed gel into layers.

The determining factor in separation of the two-component (ZnS + PbS) gel into layers in the early stage of diagenesis is the crystallization power of the components. It is well known that the simpler the type of crystal structure of a material, the greater its crystallization power. In view of this, lead sulfide should have a higher crystallization power than zinc sulfide, and the amorphous lead sulfide particles formed during coagulation of the gel will crystallize rapidly. The colloidally dispersed crystalline particles of PbS grow by collective crystallization. Under the influence of gravity, the largest settle to the lower part of the precipitate, accumulating there to form an almost monomineralic zone. Some of the finer particles do not reach the lower zone, owing to the sharply increasing viscosity of the gel precipitate, and they become suspended. By continuing to grow, these particles form dendritic skeletons in the globular zinc sulfide.

Aging of the layered ZnS + PbS gel successively in humid air and dry air was carried out for about one and one-half years.

Even at the end of the first stage of diagenesis (aging in solution) contraction cracks of exfoliation, which immediately became filled with the solution, were formed in the layered gel at the boundary between the sphalerite and sphalerite–galenite layers. Upon diagenesis of the precipitate in humid air these cracks grew sharply to 0.5 to 0.7 cm. They remained filled with the mother liquor for three weeks. It is precisely during this period that their walls gradually became clearly reniform. The most distinct reniform elements developed on the surface of the monomineralic sphalerite layer. The surface of the underlying sphalerite–galenite gel also became reniform, but with hemispheroidal elements having considerably larger radius of

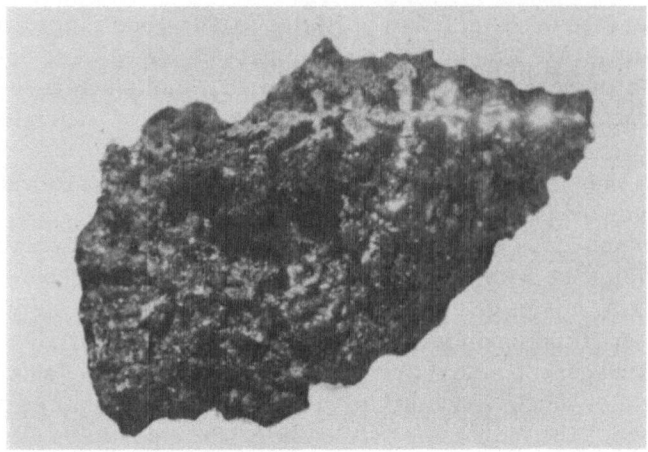

Fig. 165. Dense gelatinous mass of galenite composed of dendritic
skeletal galenite crystals (× 40).

curvature than those of the sphalerite gel. The reniform character
of the surfaces of the exfoliation cavities remained unchanged during
further aging of the differentiated precipitate.

Finally, after one and one-half years of aging, the precipitate of the
upper (sphalerite) zone was transformed into a friable powdery aggregate. The reniform surfaces were covered with very fine cracks and
were partially disintegrated. An X-ray photograph shows all the
principal sphalerite lines. The precipitate of the intermediate (sphalerite–galenite) zone also took on the consistency of a friable aggregate, but somewhat more dense than the sphalerite.

After one and one-half years of aging, the precipitate of the lower
(galenite) zone became a black, porous, and rather firm aggregate.
The structure of the spongy galenite mass could be observed clearly
under the binocular microscope at highest magnification. The
galenite mass was an accumulation of dendritic skeletal galenite
crystals (Fig. 165). These crystals formed intergrowths and mutual
interweavings of the branches of the dendrites; this caused the spongy
appearance of the aggregate.

The skeletal galenite crystals were ribbed. Distinctive cone-
shaped and cross-shaped individuals predominated in cross section
along {100}. The individual shafts of the skeletons had a scepterlike
appearance caused by some development of the {100} faces on their
terminations (Fig. 165). The size of the skeletal galenite crystals
ranged from hundredths of a millimeter to 0.7 mm. The length of the
individual shafts reached 0.3 mm.

Comparison of the galenite particles in the gel precipitate after four months of aging with those in the spongy aggregate that formed by further aging (one and one-half years) showed substantial changes in size and morphology.

On the average the particle size increased a hundredfold over this period of aging. Morphologically the crystals in the final aggregate had a distinctly skeletal appearance, in contrast to the particles forming the gel after four months of aging. This was caused primarily by rib growth of crystals and increased viscosity of the medium.

Polished sections could not be made because the material disintegrated upon sectioning, and so the spongy galenite aggregate was not studied mineragraphically.

Aging of Mixed ZnS + PbS Gels with PbS Predominant

The features of the aging of the ZnS + PbS gels in which lead sulfide is predominant are somewhat different from those described above. As before, a mixed gel with proportions PbS : ZnS = 20:1 was obtained. The precipitate occupied about three-quarters of the container immediately after it formed and was grayish black and com-

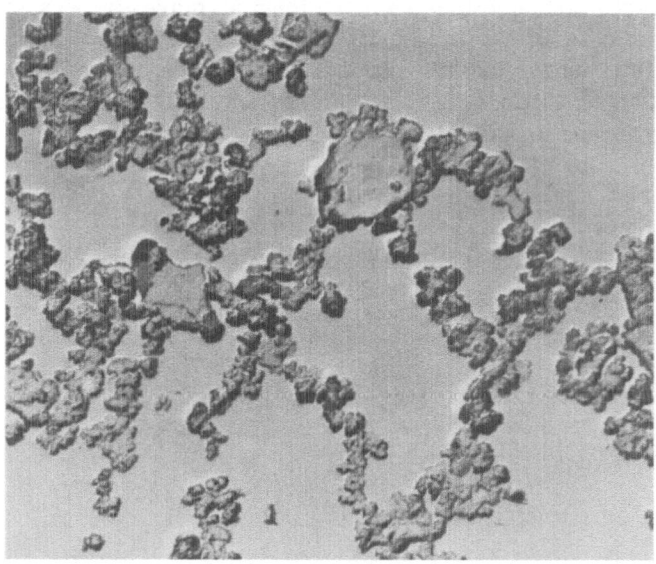

Fig. 166. Two types of particles in PbS gel aged one month. Carbon replica (× 12,000).

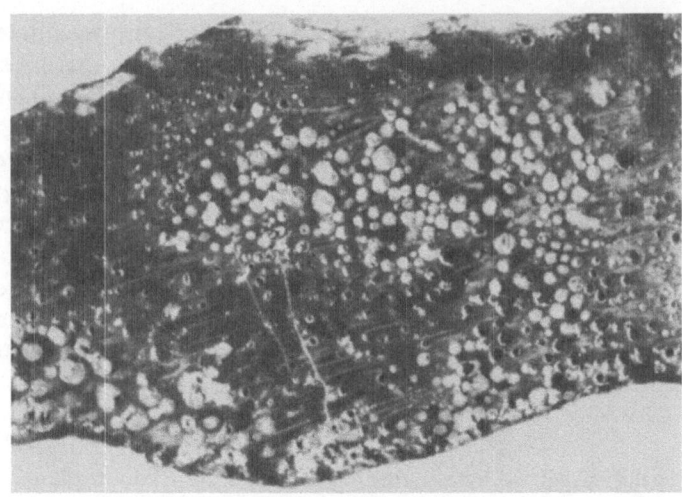

Fig. 167. Spheroidal accumulations of metacolloidal sphalerite in finely dispersed galenite (x 5).

pletely homogeneous.   Over several hours it shrank sharply, became thicker, and did not retain its uniformity of color.

On the second day of aging, the gel shrank still further and became still thicker, and spotlike segregations of more or less spheroidal white particles appeared in it.   Their distribution in the overall black gel was uniform. Over the next two to three days these spherical particles grew appreciably, reaching 0.5 to 0.8 mm in diameter. After this, there was no significant increase in their size.

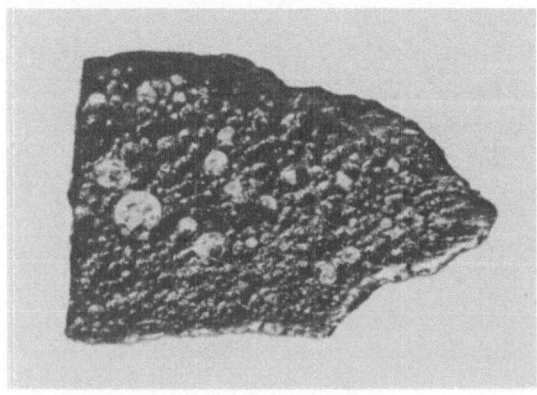

Fig. 168. Globular structure of sphalerite forming spherical segregations (x 20).

After one month of aging in the mother liquor, the precipitate was studied under an electron microscope. Colloidal ZnS particles were almost completely removed from the precipitate when the sample was prepared. The remaining PbS precipitate consisted of two kinds of crystal particles (Fig. 166). The first were isolated, relatively large, (0.7 to 1 μ), and quite rare. They were equidimensional, with clear features of skeletal growth. The second were smaller (0.05 to 0.2 μ), with a cubic or cubic-octahedral habit, and formed chain aggregates.

Aging of the precipitate in the solution was continued for four months, after which the precipitate had the consistency of a solid aggregate. Then it was held successively in humid air (two months) and in dry air. The final product was a dense black crust 6 mm thick. Very small, yellowish-white, spherical particles of zinc sulfide were evenly distributed in the main galenite mass (Fig. 167).

A uniformly globular texture of the spherical segregations of sphalerite was clearly seen in polished sections under the binocular microscope (Fig. 168). Thus, in morphology and texture these segregations were globulites. Genetically they can be considered emulsion suspensions of relatively light ZnS gel in a high-density lead sulfide gel.

The same interrelationship of components as was observed in artificially produced sphalerite–galenite aggregates are found in natural metacolloidal sphalerite–galenite aggregates. In studying the

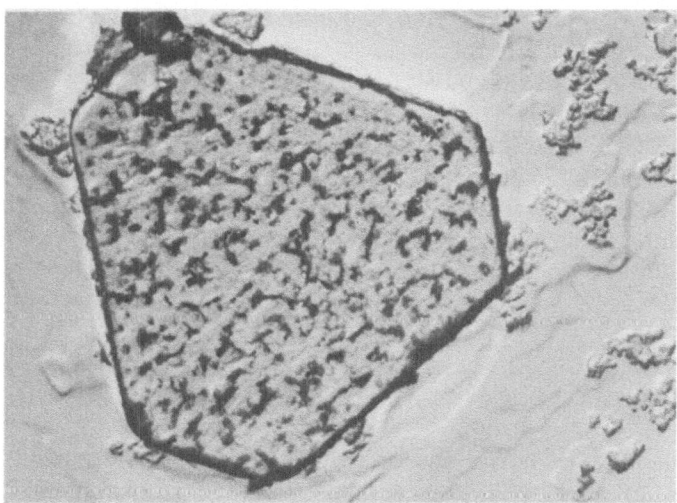

Fig. 169. Nature of particles in zinc and iron sulfide gels (with Zn : Fe = 100:10). Carbon replica (x 11,000).

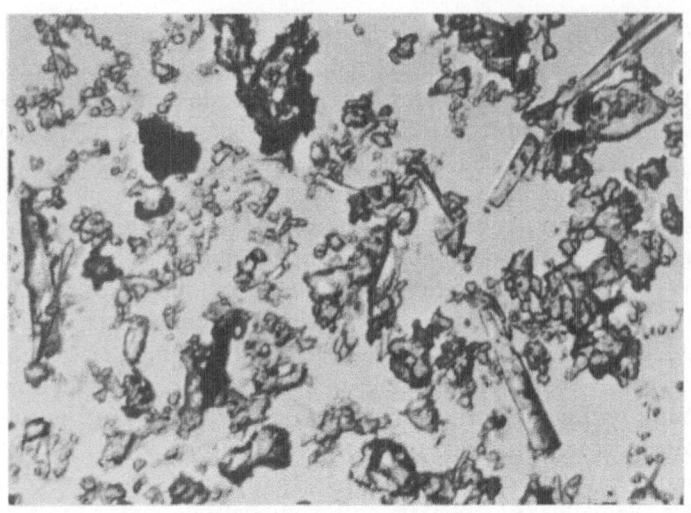

Fig.   170.   Nature of particles in zinc and iron sulfide gels (with
Zn : Fe = 100:20). Carbon replica (x 9000).

character of the galenite segregations in the metacolloidal sphalerite,
the following features are observed:

1. If sphalerite is the predominant mineral and galenite is
sharply subordinate, the galenite inclusions are skeletal forms. In the
ideal case the galenite inclusions are skeletal dendrites.

2. If galenite forms something like 20 to 40% of the metacolloidal
aggregate, it generally forms denticulated aggregates that are complex
intergrowths of skeletal dendrites.

3. If galenite is sharply predominant, it forms dense aggregates
of very small equidimensional grains in which sphalerite globulites
are more or less evenly distributed.

The clearest examples of such interrelationships of the principal
components in sphalerite–galenite metacolloidal aggregates are the
metacolloidal ores of Iokun'zh (USSR) and Olkusz (Poland).

BRIEF DATA ON FINAL AGING PRODUCTS OF MIXED ZnS + FeS GELS

It should be borne in mind that the results of studying aging
products of gels with this composition are preliminary. Gels of
ZnS + FeS composition were obtained from mixed 3% solutions of
$ZnSO_4$ and $FeSO_4$ in which the proportions of these components were

100:5, 100:10, and 100:20, respectively. Aging was in stages, as in all of the experiments described above, and totaled about one and one-half years. The exception was the precipitate obtained from a mixed solution with proportions ZnS : FeS = 100:20. To avoid oxidation of the FeS, this precipitate was held for the indicated period in a solution saturated with hydrogen sulfide. The precipitates were studied by X rays and electron microscopy.

The precipitate obtained from solutions with proportions ZnS : FeS = 100:10 was brownish yellow. Study under the electron microscope showed that this precipitate consisted of two kinds of particles. Very small equidimensional particles 0.05 to 0.1 μ in size predominated (70 to 80%). Single particles were rare; generally they formed distinctive flaky aggregates by clumping together. Other, larger particles 5 to 7 μ in size were lamellar hexagonal and ditrigonal crystals (Fig. 169). Distinct block structure on the pinacoid faces was observed. The X-ray photograph (Table 28, sample 102) corresponds to wurtzite.

Table 28. Interlayer Spacings of Precipitates of (Zn, Fe)S Composition

Line no.	Sample 101		Sample 102		Sample 103		Wurtzite	
	d	I	d	I	d	I	d	I
1	—	—	—	—	—	—	3.607	2
2	—	—	—	—	3.42	5	3.435	3
3	30.308	10	3.280		—	—	3.283	4
4	3.127	10	3.123	10	3.07	10	3.107	10
5	2.858	5	2.981	5	—	—	2.973	1
6	2.77	5	—	—	—	—	—	3
7	2.691	5	2.685	2	—	—	2.702	1
8	2.365	5	—	—	—	—	—	—
9	—	—	2.076	3	2.09	7	2.094	4
10	1.902	10	1.909	8	1.888	10	1.902	10
11	—	—	1.802	5	1.790	4	1.790	3
12	1.662	6*	1.700	3	—	—	1.645	2
13	—	—	1.628	8*	1.614	9	1.625	9
14	—	—	—	—	—	—	1.561	2
15	—	—	1.346	3	—	—	1.359	1
16	—	—	1.227	3	1.232	3	1.243	3
17	—	—	—	—	1.213	3	1.213	3
18	—	—	—	—	1.091	4	1.106	8
19	—	—	—	—	—	—	1.044	8

*Wide.

The precipitate obtained from solutions with proportions ZnS : FeS = 100:20 and held for one and one-half years in the mother liquor saturated with hydrogen sulfide was dark brown. Study under the electron microscope showed that this precipitate also consisted of two kinds of particles (Fig. 170). One kind was equidimensional, indistinctly crystallized, and 0.1 to 0.3 μ in size. The other kind was acicular and spindle-shaped, 0.4 to 3 μ in length, and 0.05 to 0.6 μ across. The X-ray photograph of this precipitate (Table 28, sample 103) corresponds to wurtzite. Thus, preliminary experiments made to obtain mixed ZnS + FeS gels show the regular appearance of the hexagonal modification of zinc sulfide at high iron concentrations in the original solutions.

The absence of lines of any other phases in the X-ray photographs is evidence that iron sulfide was minor in the precipitates and that its particles were amorphous to X-rays. Apparently much of the iron isomorphically replaced zinc in the wurtzite. The quantitative aspects of this replacement remain unclear and require further experiments. Nonetheless, it can be hypothesized that the formation of the hexagonal modification of zinc sulfide is related precisely to isomorphic replacement of zinc by iron under conditions of atmospheric pressure, low temperatures, and high concentrations of the components.

It is noteworthy that an elevated iron content is also observed in natural metacolloidal wurtzites. Wurtzites of pyrite deposits and certain other deposits are an example of this. Shadlum (1942) has pointed out the considerable iron content in wurtzite from the Yaman-Kasa deposit. According to Ehrenberg (1931), in metacolloidal zinc sulfide aggregates from the Aachen region, which consist of two polymorphic modifications, wurtzite and sphalerite, sphalerite is distinguished from wurtzite by low iron content.

BRIEF CONCLUSIONS

The foregoing experiments made it possible to note a number of patterns in the diagenetic changes in zinc sulfide gels and two-component (ZnS + PbS) gels.

1. Immediately after the zinc sulfide gel coagulated, it began to globulate, expressed as the separation of the overall structureless mass of the gel into numerous lumps, at first friable, which 10 to 12 minutes after precipitation became more distinctly spherical, with small, very dense envelopes. The presence of NaCl and CaCl$_2$ in the solution accelerated globulation of the precipitate. Globulation was

caused by surface-tension forces, with the formation of spherical shapes, as those with the smallest surface per unit volume, requiring the smallest surface energy.

2. During aging of the globulated gels in the first stage of diagenesis (aging in solution), several globules became attached to form spherical aggregates (globulites). The longer the first stage of diagenesis of the precipitate and the higher the concentration of NaCl and $CaCl_2$ in the solution above the precipitate, the more abundantly globulites formed in the main globular mass. The formation of globulites also corresponded to a general tendency of the material to decrease its surface energy.

3. The formation of oolites in these experiments was caused by repeated precipitation of zinc sulfide with deposition of the newly formed gel on the surface of the globulites and, less often, on individual globules, which in this case were centers of oolite formation.

4. The reniform surfaces formed in the contraction cracks were not related to crystallization processes, because they were still formed in viscous gels.

5. Crystallization of the ZnS gels began after five to six months' aging in dry air. In the presence of small admixtures of $CaCl_2$ and $MgCl_2$ (0.1 to 0.5%) in the medium, the gels crystallized in the early stage of aging (in the solution) one month after they were formed. A small admixture of chlorides in dry ZnS precipitates also intensified recrystallization and caused the formation of coarser aggregates.

6. High calcium chloride concentration in the original solutions, leading during diagenesis of the mixed gel to the formation of gypsum, sharply decelerated crystallization of the ZnS gel and was thus a distinctive stabilizer of primary textures.

7. Separation of the two-component ZnS + PbS gel into layers was caused by the differing crystallization power of the components and was expressed either as gravitational differentiation of the precipitiate or as collective crystallization of particles of one component suspended in the mass of the other.

8. Crystallization of galenite in the zinc sulfide gel had a skeletal character.

9. The textures and structures of metacolloidal sphalerite-galenite aggregates produced in the laboratory were similar to those of natural aggregates and were a criterion of the colloidal origin of these natural aggregates.

Chapter 10

Mineralogy of the Pauzhetka Natural Steam Deposit

Modern processes of mineral formation always attract the investigator by the alluring prospect of being able to observe the process directly.

In the outlet region of the Pauzhetka hot springs there is intense hydrothermal metamorphism of tuffaceous rocks, and, though the overall features of this metamorphism have been studied to some extent, mineralogical studies are still amost in their infancy. The data presented here correspond precisely to this initial stage of the study of the deposits.

Study of the considerable accumulations of silica gel in the hot springs and creeks seems to be of special interest in addition to the overall features of mineral formation.

BRIEF GEOLOGIC AND PETROGRAPHIC DESCRIPTION

The Pauzhetka springs, a strong outlet of thermal waters, is in southern Kamchatka close to Koshelev and Kambal'nyi volcanoes. The hot springs are restricted to a trough with roughly north-south strike which was formed at the beginning of the Quaternary. Development of the region of trough subsidence in Quaternary time led ultimately to sharp differentiation: the formation of the Kambal'nyi Range anticlinical structure and relict basins occupied by the valleys of the Pauzhetka and Ozernaya rivers.

Efflux of the thermal waters is caused by the low elevation of the outlet (100 m) and tectonic crushing of the western limb of the Kambal'nyi Range anticline. The outflux patterns of the Pauzhetka hot springs were studied in detail by Aver'ev (1961), who on the basis of drilling data and detailed geological study of the region developed the picture presented in Fig. 171. The springs discharge from Middle Quaternary agglomeratic tuffs that form thick flows (about 200 m). According to Aver'ev, the outlet of the Pauzhetka hot springs is a considerable distance from possible sources of heating. He supposed that the waters are strongly heated deep in the Kambal'nyi Range.

The thermal waters come out to the surface mainly along a large fracture and a well-worked channel. The water temperature is 90 to 100° at the outlets but increases considerably with depth, reaching a maximum to 170 to 180° at 120 to 350 m (well no. R-1).

The Pauzhetka hot-spring waters are on the whole alkaline and contain sodium chloride, though there is some decrease in pH with depth. Thus, Ivanov (1961) found a pH of 8.6 at the outlet of Pul' siruyushchii Spring. Aver'ev (1961) noted a pH of 8 for solutions at depths of 120 to 196 m. The mineral content of the waters is about 3 g/l. A substantial silica content, reaching almost 0.2 g/l, is characteristic. Table 29 gives the chemical composition of the Pauzhetka waters at various depths. Condensed steam of the thermal solutions contains chemically active gases (CO_2, H_2S, NH_3) and is acid, with a pH of 5.5. The sequence of volcanic and pyroclastic rocks filling the Quaternary trough has been hydrothermally metamorphosed by the Pauzhetka hot-spring waters in the outlet region. A rotary well drilled to 800 m exposed the following hydrothermally altered rocks:

Rocks	Depth interval, m
Peplitic dacite tuffs	5—60
Agglomeratic lithic dacite tuffs	30—165
Lithic dacite tuffs with andesite and andesite—basalt fragments of various sizes	165—263
Agglomeratic andesite—basalt tuffs	263—382
Vitric-crystal dacite tuffs, very dense, consisting of fused fragments of quartz, hornblende, and fragments of andesite and andesite—basalt lava up to 5 mm in size. Total quantity of lava fragments less than 5%	382—570
Agglomeratic andesite tuff with fragments of andesite—basalt	570—657
Dacite and andesite tuffs (grain size ranging from silt to fine-gravel size) interbedded with pebble conglomerates	657—800

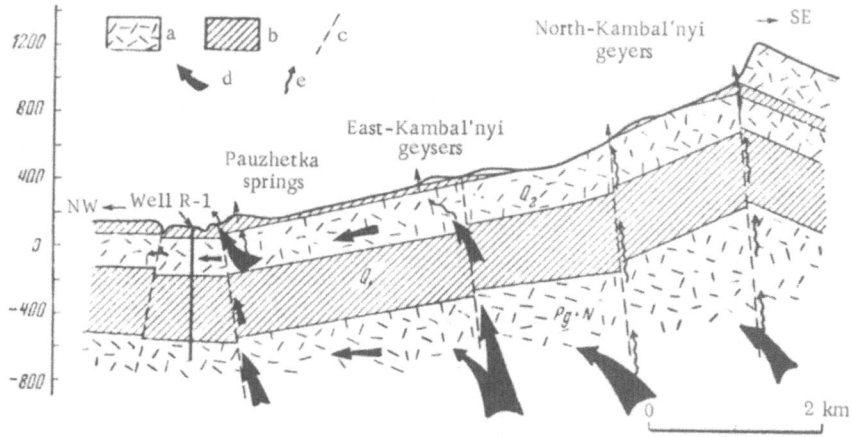

Fig. 171. Diagram of heating of the Pauzhetka hot springs (after V. V. Aver'ev). (a) Rocks supplied with water; (b) relatively impermeable rocks; (c) faults; (d) water; (e) steam.

Naboko (1961, 1963; Naboko and Lebedev, 1963; and Naboko and Piip, 1961) has studied the recent hydrothermal metamorphism of the Pauzhetka tuffs. He established by geologic and petrographic work on the hydrothermally altered rocks that the hydrothermal meta-morphism at Pauzhetka involves propylitization, zeolitization, and kaolinization of the pyroclastics. He also clarified the vertical metasomatic zonation of these processes and briefly described the mineral alteration products typical of each zone.

According to Naboko, the kaolinized zone extends from the surface to a depth of 5 m. The kaolinized rocks have been altered to clays which often preserve the textures of the original rocks.

The patterns of the alteration of the tuffs below the kaolinized zone have been studied very little. Taking into account the alkalinity of the thermal solutions and the small outflux of bases into the upper layers of the zeolitized zone, it can be hypothesized that between the latter and the kaolinized rocks is a zone of alteration to montmor-illonite. Noll (1936) established experimentally that montmorillonite forms under alkaline conditions when the proportions of reacting components are RO $(R_2O) : Al_2O_3 : SiO_2 = 0.2 : 1 : 4$. Increases in the concentration of NaOH to the proportions $Na_2O : Al_2O_3 : SiO_2 = 1 : 1 : 4$ lead to the formation of analcime. In the actual conditions of altera-tion of the Pauzhetka tuffaceous rocks there is intense outflux of sodium and also some regrouping of calcium, with noticeable transfer of calcium from the lower to upper layers of the altered rocks. Under

Table 29. Chemical Composition of Pauzhetka Hot-Spring Waters*

Component	1 Pul'siruyushchii spring			2 Well R-1 interval 120-196 m			3 Well (?) interval 300-400 m		
	g/l	mg/equiv.	equiv. %	g/l	mg/equiv.	equiv. %	g/l	mg/equiv.	equiv. %
Cations									
NH_4^+	Not found	–	–	0.0004	0.02	0.04	0.0007	–	–
Na^+	0.9486	41.26	90.1	1.0540	45.85	94.45	0.9398	–	–
K^+	0.0648	1.65	3.6	0.0070	0.57	1.17	0.0070	–	–
Mg^{2+}	0.0037	0.30	0.6					–	–
Ca^{2+}	0.0519	2.59	5.7	0.0422	2.11	4.34	0.1192	–	–
Sum	1.0693	45.80	100.0	1.1036	48.55	100.0	1.0667	–	–
Anions									
F^-	0.0009	0.04	0.1	–	–	–	–	–	–
Cl^-	1.5377	43.36	94.4	1.6380	46.19	95.15	1.4700	–	–
Br^-	0.0042	0.05	0.1	–	–	–	–	–	–
I^-	Not found	–	–	–	–	–	–	–	–
SO_4^{2-}	0.0727	1.51	3.3	0.0654	1.77	3.64	0.1642	–	–
HCO_3^-	0.0335	0.55	1.1	0.0360	0.59	1.21	0.0121	–	–
CO_3^{2-}	0.0090	0.30	0.7	–	–	–	0.0708	–	–
$HSiO_3^-$	0.0078	0.10	0.3	–	–	–	–	–	–
Sum	1.6658	55.91	100.0	1.7594	48.55	100.0	1.7171	–	–
Undissociated molecules H_2SiO_3	0.204	–	–	0.1990	–	–	0.1703	–	–
Total mineralization	3.05	–	–	3.2	–	–	3.07	–	–
pH	8.2			8.0				–	–

*Notation: 1, data of V.V. Ivanov (1961); 2, data of V.V. Aver'ev (1961); 3, data of S.I. Naboko (1961).

these conditions, with molecular proportions $RO(R_2O) : Al_2O_3 : SiO_2$ = 1:1:4, with replacement of sodium by calcium, wairakite can form. Apparently Noll also obtained wairakite in his experiments; he noted it as a material difficult to identify, which is quite natural because wairakite was first described as a new mineral only in 1955 (Steiner, 1955).

It should be noted that calcic montmorillonite will form in the zone of alteration to montmorillonite at elevated calcium concentrations.

Table 30. Chemical Composition of Zeolitized
Dacite Tuffs
(Horizon 42.4 to 50.7 m; Well R-1)

Component	%	Atomic proportions	Atomic proportions of oxygen	Number of cations
SiO_2	60.89	1014	2028	538
TiO_2	0.67	009	18	5
Al_2O_3	15.14	296	444	157
Fe_2O_3	2.33	028	42	15
FeO	2.52	035	35	18
MnO	0.09	—	—	—
MgO	1.83	045	45	24
CaO	4.33	077	77	41
Na_2O	2.06	064	32	34
K_2O	4.55	098	49	52
H_2O^+	3.93	434	217	230
H_2O^-	0.61	—	—	—
CO_2	0.77	018	36	10
S	0.30	009	12	—
Sum. . . .	100.02		3011	1124

$1600 : 3011 = 0.531$

Unaltered dacite:

$$K_{29}Na_{70}Ca_{34}Mg_{19}Fe^{2}_{14}Fe^{3}_{16}Al_{156}Ti_3Si_{616}O_{1600}$$

Zeolitized dacite tuff (42-50 m):

$$K_{52}Na_{34}Ca_{41}Mg_{24}Fe^{2}_{18}Fe^{3}_{15}Al_{157}Ti_5Si_{538}C_{10}O_{1361}(OH)_{230}S_9$$

Influx of ions		Outflux of ions
K — 23	Al — 1	Si — 72
Ca — 7	Ti — 2	Na — 36
Mg — 5	C — 10	
Fe — 3	OH — 230	
	S — 9	

Table 31. Chemical Composition of Zeolitized
Dacite Tuff
(Horizon 103–110 m; Well R-1)

Component	%	Atomic proportions	Atomic proportions of oxygen	Number of cations in standard cell
SiO_2	60.32	1004	2008	529
TiO_2	0.67	009	018	5
Al_2O_3	15.22	298	447	157
Fe_2O_3	2.74	034	051	18
FeO	2.56	035	035	18
MnO	0.11	—	—	
MgO	2.28	057	057	20
CaO	4.91	087	087	46
Na_2O	2.45	080	040	42
K_2O	2.43	050	025	26
H_2O^+	4.53	500	250	263
H_2O^-	0.75	—	—	—
CO_2	0.66	015	030	8
S	0.33	010	014	
Sum. . . .	99.96		3034	1132

$$1600 : 3034 = 0.527$$

Unaltered dacite:

$$K_{29}Na_{70}Ca_{34}Mg_{19}Fe^2_{14}Fe^3_{16}Al_{156}Ti_3Si_{610}O_{1600}$$

Zeolitized dacite tuff (103–110 m):

$$K_{26}Na_{42}Ca_{46}Mg_{20}Fe^2_{18}Fe^3_{18}Al_{157}Ti_5Si_{529}C_8O_{1327}(OH)_{263}S_{10}$$

Influx of ions		Outflux of ions
Ca — 12	Al — 1	Si — 81
Mg — 1	Ti — 2	Na — 28
Fe — 6	C — 8	K — 3
S — 10	OH — 263	

This hypothesis is confirmed by the discovery of reworked calcic montmorillonite in cementing minerals in alluvium.

The zeolitized zone lies from 30 to 250 m. Mainly agglomeratic and lithic dacite tuffs and to a lesser extent peplitic dacite tuffs have been hydrothermally decomposed and altered here. Glass, which in most of the rock is completely replaced by finely crystalline laumontite, has been the most strongly altered. Feldspars, mainly intermediate and less commonly acidic plagioclases, have been altered almost as strongly; they are almost completely replaced by laumontite and less commonly are leached completely.   Wedge-shaped adularia crystals are in places present in the leaching cavities. Authigenic adularia is most common in altered rocks from 50 to 85 m. The alteration of inclusions of dark components, mainly augite, less often hornblende, has been extremely minor. In regions of the rock enriched in pyrite, the pyrite sometimes pseudomorphically replaces augite. Augite and

Table 32. Chemical Composition of Zeolitized
Dacite Tuff
(Horizon 220-236 m; Well R-1)

Component	%	Atomic proportions	Atomic proportions of oxygen	Number of cations in standard cell
SiO_2	58.96	981	1962	514
TiO_2	0.84	010	20	5
Al_2O_3	16.46	324	486	170
Fe_2O_3	2.75	034	51	18
FeO	2.51	035	35	18
MnO	0.14	002	2	1
MgO	2.45	062	62	32
CaO	5.12	091	91	48
Na_2O	2.50	080	40	42
K_2O	1.99	042	21	22
H_2O^+	5.18	578	289	302
H_2O^-	0.81	—	—	—
CO_2	0.10	002	4	1
S	0.15	005	7	—
Sum. . . .	99.96		3056	1172

$$1600 : 3056 = 0.524$$

Unaltered dacite:

$$K_{29}Na_{70}Ca_{34}Mg_{19}Fe^2_{14}Fe^3_{16}Al_{156}Ti_3Si_{610}O_{1600}$$

Zeolitized dacite tuff (220-236 m):

$$K_{22}Na_{42}Ca_{48}Mg_{32}Fe^2_{18}Mn_1Fe^3_{18}Al_{170}Ti_5Si_{514}C_1O_{1293}(OH)_{302}S_5$$

Influx of ions		Outflux of ions
Ca — 14	Fe — 6	Si — 96
Mg — 13	Al — 14	Na — 28
Mn — 1	Ti — 2	K — 7
S — 5	OH — 302	
	C — 1	

Table 33. Chemical Composition of Dikii Ridge Dacite

Component	%	Atomic proportions	Atomic proportions of oxygen	Number of cations in standard cell
SiO_2	68.50	1140	2280	610
TiO_2	0.39	005	010	3
Al_2O_3	14.91	292	438	156
Fe_2O_2	2.42	030	045	16
FeO	1.88	026	026	14
MnO	—	—	—	—
MgO	1.36	035	035	19
CaO	3.58	064	064	34
Na_2O	3.98	130	065	70
K_2O	2.46	054	027	29
H_2O	0.42	—	—	—
Sum	99.90		2990	951

$$1600 : 2990 = 0.535$$

Formula of rock:

$$K_{29}Na_{70}Ca_{34}Mg_{19}Fe^2_{14}Fe^3_{16}Al_{156}Ti_3Si_{610}O_{1600}$$

hornblende are slightly chloritized in the deeper layers of the zeolitized zone. The accessory ore minerals, hematite and ilmenite, have been altered appreciably. Hematite is completely altered to pseudomorphic magnetite, and the new magnetite is redeposited on a considerable scale as octahedral crystals on the walls of cavities formed by the leaching of feldspars. Ilmenite is replaced by finely crystalline aggregates of anatase and brookite. Upon more intense decomposition and solution, in the deeper layers of the zone, ilmenite is replaced by sphene.

Chemical analyses of zeolitized tuffs from various layers of the zeolitized zone, taken from Naboko (1963), have been recalculated by Barth's method to clarify the features of the transfer of components during zeolitization of the agglomeratic lithic tuffs (Tables 30 to 32). The results are compared with the formula of unaltered dacite of Dikii Ridge (Table 33). Even though there are no data on the porosity of the rocks, the results of the recalculations reflect the main patterns of transfer of the principal components of the rocks. Small variations in the influx and outflux of individual components are not taken into account: Transfer values less than 10 are neglected. We obtain the following orders of influx and outflux of components during zeolitization of the dacite tuffs at various depths:

42–50 m

Influx of ions		Outflux of ions
OH–230	C–10	Si–72
K– 23	S– 9	Na–36

There is also a distinct tendency toward influx of Ca.

103–110 m

Influx of ions		Outflux of ions
OH–263	S–10	Si–81
Ca– 12	C– 8	Na–28

220–236 m

Influx of ions		Outflux of ions
OH–302	Ca–14	Si–96
Al– 14	Mg–13	Na–28

There is also a tendency toward influx of S and outflux of K.

Thus, zeolitization of the dacite tuffs involved distinctive hydration and the formation of zeolites. In connection with the strong outflux of sodium, these zeolites are generally calcic varieties: calcic laumontite, heulandite, wairakite, stellerite, and (much less common) mordenite and thomsonite. Zeolitization of the dacite tuffs involved considerable outflux of silica and, in connection with this, a substantial enrichment of the solutions in this component. Apparently a considerable part of the silica in the Pauzhetka thermal solutions owes its origin to this process.

The propylitized zone lies from 250 to 800 m. Agglomeratic andesite and andesite–basalt tuffs and also vitric and lithic dacite tuffs have been altered. Glass and dark minerals have been most strongly altered in the propylitization of the andesite and andesite–basalt tuffs. Glass fragments are strongly replaced by aggregates of finely dispersed chlorite. Of the original dark minerals only titanaugite has been reliably identified ($+2V = 52°$; $a = 1.716 - 1.718$; $\beta = 1.722 - 1.724$; $\gamma = 1.743 - 1.748$. It is noteworthy that fresh titanaugite is exceedingly rare; generally the mineral is completely replaced by an aggregate of chlorite, gyrolite, and sphene. The presence of such pseudomorphism after titanaugite is characteristic of the altered rocks of the propylitized zone. Also, rocks in the upper parts of this zone contain abundant small cavities filled with clear blue calcic chlorite (erinite). Feldspars, mainly intermediate plagioclases, also have been considerably altered and are replaced by calcite and

anhydrite in the upper layers and calcite in the lower layers of propylitized zone. Ilmenite is replaced by anatase and brookite. In the
upper parts of the zone, ilmenite is often completely leached; there
are relatively large crystals of anatase and brookite in the leaching
cavities and spherulites of sphene in the surrounding chloritized rock.
Hematite is replaced by magnetite and pyrite. Pyrite, which together with hematite replaces the dark mineral components on a large
scale, is abundantly disseminated in the rock.

Table 34. Chemical Composition of Propylitized
Andesite Tuff
(Horizon 320 m; Well R-1)

Component	%	Atomic pro-portions	Atomic pro-portions of oxygen	Number of cations in standard cell
SiO_2	51.30	854	1708	474
TiO_2	0.93	011	22	6
Al_2O_3	15.38	302	453	167
Fe_2O_3	5.00	062	93	34
FeO	4.60	064	64	35
MnO	—	—	—	—
MgO	2.50	062	62	34
CaO	6.12	109	109	60
Na_2O	1.10	036	18	20
K_2O	1.11	024	12	13
H_2O^+	6.33	700	350	388
CO_2	2.27	051	102	28
S	2.54	079	110	
Sum	99.18		2883	1259

$$1600 : 2883 = 0.555$$

Unaltered andesite:

$$K_8Na_{53}Ca_{100}Mg_{57}Fe_{35}^2Fe_{40}^3Al_{209}Ti_7Si_{473}O_{1600}$$

Propylitized andesite tuff (320 m):

$$K_{13}Na_{20}Ca_{60}Mg_{34}Fe_{35}^2Fe_{34}^3Al_{167}Ti_6Si_{474}O_{1105}C_{28}(OH)_{388}S_{79}$$

Influx of ions		Outflux of ions	
K — 5	OH — 388	Na — 33	Fe — 6
Si — 1	S — 79	Ca — 40	Al — 42
C — 28		Mg — 23	Ti — 1

Table 35. Chemical Composition of Dacite Tuff
from Propylitized
(Horizon 510 m; Well R-1)

Component	%	Atomic proportions	Atomic proportions of oxygen	Number of cations in standard cell
SiO_2	68.64	1142	2284	595
TiO_2	0.34	004	8	2
Al_2O_3	15.55	305	457	159
Fe_2O_3	1.55	020	30	10
FeO	1.90	026	26	13
MnO	0.10	—	—	
MgO	1.50	037	37	19
CaO	3.76	066	66	34
Na_2O	2.83	090	45	47
K_2O	2.43	050	25	26
H_2O^+	1.45	166	83	43
H_2O^-	0.18	—	—	—
CO_2	0.15	003	6	2
P_2O_5	Trace	—	—	—
Sum	100.38		3067	950

$1600 : 3067 = 0.521$

Unaltered Dikii Ridge dacite:

$$K_{29}Na_{70}Ca_{34}Mg_{19}Fe^2_{14}Fe^3_{16}Al_{156}Ti_3Si_{610}O_{1600}$$

Weakly altered dacite tuff (510 m):

$$K_{26}Na_{47}Ca_{34}Mg_{19}Fe^2_{13}Fe^3_{10}Al_{159}Ti_2Si_{595}C_2O_{1557}(OH)_{43}$$

Influx of ions		Outflux of ions		
Al — 3	OH — 43	K — 3	Fe — 7	Si — 15
C — 2		Na — 23	Ti — 1	

Also there is a noticeable tendency toward outflux of titanium.

The vitric-crystal dacite tuffs in the propylitized zone are little altered. Glass and feldspars are weakly chloritized.

As with the zeolitized rocks, chemical analyses of the altered tuffs of the propylitized zone from the work of Naboko (1963) have been recalculated by the writer. The results of the recalculation (Tables 34 to 37) are compared with the formulas of unaltered dacite from Dikii Ridge andesite from the Kambal'nyi Range. As before, influx and outflux values less than 10 are neglected. We obtain the following main orders of influx and outflux of components during

Table 36. Chemical Composition of Propylitized Andesite Tuff
(Horizon 620 m; Well R-1)

Component	%	Atomic proportions	Atomic proportions of oxygen	Number of cations in standard cell
SiO_2	54.00	899	1798	494
TiO_2	0.35	005	10	3
Al_2O_3	14.16	277	415	152
Fe_2O_3	2.10	026	39	14
FeO	2.51	035	35	19
MnO	0.17	003	3	2
MgO	3.34	082	82	45
CaO	9.25	166	166	91
Na_2O	2.48	080	40	44
K_2O	2.47	054	27	29
$H_2O\pm$	1.07	122	61	67
CO_2	6.55	148	296	81
S	1.70	053	74	—
Sum	100.15		2898	

$$1600 : 2898 = 0.550$$

Unaltered andesite:

$$K_8Na_{53}Ca_{100}Mg_{57}Fe^2_{35}Fe^3_{40}Al_{209}Ti_7Si_{473}O_{1600}$$

Propylitized andesite tuff (630 m):

$$K_{29}Na_{44}Ca_{91}Mg_{45}Mn_2Fe^2_{19}Fe^3_{14}Al_{152}Ti_3Si_{494}C_{81}O_{1480}(OH)_{67}S_{53}$$

Influx of ions		Outflux of ions	
K — 21	C — 81	Na — 9	Fe — 32
Mn — 2	OH — 67	Ca — 9	Al — 57
Si —21	S — 53	Mg — 12	Ti — 4

propylitization of the adesite tuffs and vitric-crystal dacite tuffs at various depths:

320 m

Influx of ions		Outflux of ions	
OH — 388	C — 28	Al — 42	Na — 33
S — 79		Ca — 40	Mg — 23

There is a tendency toward influx of potassium.

510 m

Influx of ions	Outflux of ions
OH — 43	Na — 23
	Si — 15

630 m

Influx of ions		Outflux of ions
C — 81	Si — 21	Al — 57
OH — 67	K — 21	Fe — 32
S — 53		Mg — 12

Thus, the recalculated data show that during propylitization the andesite and andesite—basalt tuffs have been strongly hydrated, with the formation of chlorites (delessite and erinite). Erinite is most widespread in the boundary zone between the propylitized and zeolitized zones.

The considerable influx of carbon dioxide and sulfur into the rock causes widespread carbonatization of the feldspars (particularly in the deeper parts of the zone) and pyritization of the hematite and the dark minerals.

The considerable outflux of aluminum, calcium, magnesium, and sodium should be noted. The first three components are carried into the upper layers and are partially shed by solutions in the boundary zone and in the zeolitized zone. A certain part of these components remain in the solutions until outflux at the surface, when they are

Table 37. Chemical Composition of Fresh
Andesite from Kambal'nyi Range

Component	%	Atomic proportions	Atomic proportions of oxygen	Number of cations in standard cell
SiO_2	50.73	844	1688	473
TiO_2	0.99	013	26	7
Al_2O_3	19.10	374	561	209
Fe_2O_3	5.65	072	108	40
FeO	4.52	063	63	35
MnO	0.14	—	—	—
MgO	4.14	102	102	57
CaO	10.00	178	178	100
Na_2O	2.89	094	47	53
K_2O	0.68	014	7	8
$H_2O^{\pm}$	1.43	156	78	87
Sum	100.27		2858	1069

$1600 : 2858 = 0.560$

Formula of rock:

$$K_8Na_{53}Ca_{100}Mg_{57}Fe^2_{35}Fe^3_{40}Al_{209}Ti_7Si_{473}O_{1600}$$

Table 38. Pauzhetka Minerals

Mineral	Distribution	Mineral	Distribution
Pyrite	++++	Sphene	++
Chalcopyrite	+	Gyrolite	++
Sphalerite	+	Prehnite	++
Hematite	+++	Chrysocolla*	++
Ilmenite	+++	Datolite	+
Magnetite	+++	Wairakite	+++
Mushketovite†	+++	Heulandite	++
Anatase	++	Laumontite	++++
Brookite	++	Stilbite	++
Rutile	++	Mordenite	++
Quartz	++++	Thomsonite	+
Chalcedony	++++	Calcite	++++
Opal (sinter)	++++	Ankerite	++
Silica gel	++++	Siderite	+
Cuprite*	+	Anhydrite	+++
Adularia	+++	Apatite	+
Albite	+++	Ca montmorillonite	+++
Delessite	++++	Kaolinite	++++
Fe erinite	+++	Hydrohematite	++++
Celadonite	++++		

++++, very widespread; +++, widespread; ++, rare; +, very rare.
*Nonnatural mineral.
†Pseudomorphs of magnetite after hematite [transl.].

precipitated together with silica gel as laumontite, adularia, and other minerals.

MINERALOGY

Study of the features of modern mineral formation at Pauzhetka makes it possible to distinguish three main genetic complexes of new minerals, caused on the whole by a single process of discharge of hot-spring solutions and related decomposition and alteration of the tuffs. The differences among these complexes lie in the conditions and mechanisms of formation of the mineral associations. The following mineral associations are distinguished:

1. Minerals formed in the hydrothermally altered rocks.
2. Minerals precipitated from solutions (and perhaps also condensed steam) as cement in alluvial deposits (and called cementation minerals in the following text).

3. Sedimentation and flow-deposited minerals of hot springs and creeks.

At present, 39 minerals are known from Pauzhetka (Table 38). Of these, 34 mineral species and varieties are genetically related to present hot-spring activity. Two minerals, chrysocolla and cuprite, owe their origin both to hot-spring waters and to the technological activity of man. The chalcopyrite was observed under the microscope at highest magnifications as an emulsion dissemination in pyrite, in only one sample. Minerals of the oxide, silicate, and carbonate groups are most widespread: opal, chalcedony, quartz, laumontite, chlorite minerals, and calcite.

Description of the Minerals

In the description of the Pauzhetka minerals, the colloidal and metacolloidal minerals are described in the most detail. Short descriptions of all the other mineral species are given.

Sulfides

Pyrite. Pyrite is a widespread mineral among the hydrothermally altered rocks of Pauzhetka. It is most widespread in the propylitized zone, where it forms an abundant and relatively uniform dissemination in the altered tuffs. The larger pyrite crystals always show skeletal growth and generally have the outlines of pyritohedra.

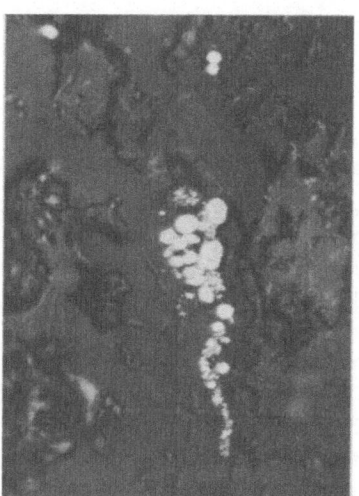

Fig. 172. Globular segregations of pyrite in voids of propylitized tuffs (deep zone) (× 500).

Table 39.   Interlayer Spacings of Sphalerite and Pyrite
(Intergrown)

Line no.	d	I	Phase	Line no.	d	I	Phase
1	3.445	4	s	16	1.349	3	s
2	3.112	10	s	17	1.241	7	s
3	2.970	1	s	18	1.214	1	s
4	2.827	2		19	1.209	2	p
5	2.698	5	p	20	1.179	2	p
6	2.406	5	p	21	1.153	1	p
7	2.206	4	p	22	1.144	1	s
8	2.108	4	s	23	1.103	9	s
9	1.917	10	s	24	1.053	2	s
10	1.795	4	s	25	1.041	9	p
11	1.629	10	p	26	1.040	3	s
12	1.560	2	p	27	1.005	4	s
13	1.498	2	p	28	1.003	3	s
14	1.442	3	p	29	0.988	4	s
15	1.363	1	s	30	0.936	3	s

s, Sphalerite lines; p, pyrite lines.

Spectrographically, only moderate titanium lines and very weak copper and manganese lines were noted in pyrite separated from the rock. The elevated titanium concentration in the pyrite is evidence that pyrite in rocks of this zone apparently formed by reworking of finely dispersed primary ilmenite.

Minor latticework zones of pyrite formed by linearly oriented capillary veinlets are less common the propylitized rocks. Moderate copper lines, weak titanium, manganese, and zinc lines, and traces of lead and antimony lines were identified spectrographically in pyrite from these veinlets.

Distinctive forms of pyrite were found in the lower part of the propylitized zone (500 to 800 m). As in the upper part of this zone, a high content of pyrite is observed here, segregated in the vitric-crystal dacite tuffs as authigenic crystals and fine-grained aggregates pseudomorphic after hematite and in Tertiary tuffites as irregularly collomorphic and spherical forms. The collomorphic forms of pyrite are lumpy reniform and vermicular aggregates and are generally restricted to leaching cavities. The spherical aggregates are of two textural-morphological types, globules and globulites (Lebedev, 1960).

The smallest spherical bodies of pyrite, i.e., globules, are observed under the microscope only at highest magnifications. Their distribution in the altered rock is uneven. Globulites, spherical aggregates of pyrite globules, are more evenly distributed but often form accumulations in cavities (Fig. 172). In most cases the pyrite globulites have uniformly globular texture; chain globular textures in the globulites are less common.

Pyrite is present everywhere in the zeolitized zone but is very unevenly distributed. It usually forms small cubic crystals. Skeletal forms are common. Pseudomorphs of pyrite after pyroxene were observed in isolated cases. Sometimes pyrite is present on laumontite faces as intergrowths of very small crystals. In isolated cases it forms continuous linings on laumontite crystals.

Sphalerite. Sphalerite is present along with pyrite in the latticework zones in the propylitized rocks. It forms very small (0.01 to 0.02 mm), irregular segregations in masses of fine-grained pyrite. Segregations of sphalerite 0.5 to 0.8 mm in size are noted in isolated cases. Under the microscope in reflected light the sphalerite is light gray and isotropic. Clear yellow internal reflections are characteristic in immersion media. Regions enriched in sphalerite were separated from a sample of pyrite containing large sphalerite inclusions. An X-ray picture of the mixture shows all the principal sphalerite lines as well as pyrite lines, here listed in Table 39.

Chalcopyrite. Chalcopyrite is found in pyrite from the lattice-work zones. It forms a very fine emulsion in the pyrite grains.

Oxides

Oxides are the most widespread minerals after silicates. The primary accessory ore minerals of the tuffs are described briefly in addition to minerals formed by the hot-spring waters.

Table 40. Interlayer Spacings of Ilmenite

Line no.	d	I	Line no.	d	I
1	3.676	5	13	1.631	3
2	3.001	3	14	1.500	7
3	2.711	10	15	1.465	9
4	2.516	7—8	16	1.340	8
5	2.455	1	17	1.266	7
6	2.341	1	18	1.209	1
7	2.214	6	19	1.185	6
8	2.046	2	20	1.153	6
9	1.880	1	21	1.117	6
10	1.856	7	22	1.076	6
11	1.716	10	23	1.050	5
12	1.649	1			

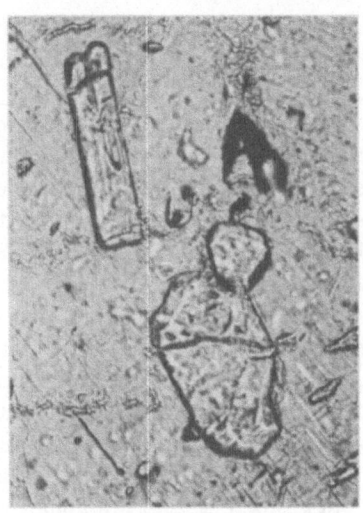

Fig. 173. Anatase and brookite crystals around altered ilmenite (upper
layers of propylitized zone) (x 120;
uncrossed nicols).

Hematite and ilmenite are the primary ore minerals of the tuffs.
The importance of magnetite among the ore minerals is unclear because of widespread formation of pseudomorphs of magnetite after
hematite.   Hematite and ilmenite are more or less evenly distributed
in the rock as small insets (approximately 0.01 mm). Larger forms
0.2 to 0.5 mm in size show a bunched distribution. In addition to being
present as individual crystals, these minerals often form mutual intergrowths.

Hematite.   Hematite is found as flattened, rounded forms, less
commonly as relatively sharply bounded crystals of thinly tabular
aspect along {0001}.   Rosette growths of lamellar crystals are even
less common. Under the microscope in reflected light the hematite is
sharply anisotropic and weakly pleochroic.   The reflection is about

Table 41. Interlayer Spacings of Anatase

Line no.	d	I	Line no.	d	I
1	3.84	5	7	1.661	6
2	3.49	10	8	1.477	7
3	2.377	5	9	1.365	2
4	2.070	3	10	1.342	3
5	1.883	8—9	11	1.262	6
6	1.694	6	12	1.166	3—4

25%. In immersion media the hematite has a distinct bluish-red internal reflection.

Ilmenite. Ilmenite is generally found as lamellar crystals, slightly rounded but always retaining elements of faces. The habit is tabular along {0001}. Under the microscope in reflected light the ilmenite is anisotropic. The color is gray, and the reflection is about 17 to 18%. Polysynthetic twinning is common. Sometimes emulsion segregations of hematite are found. Table 40 gives X-ray data for ilmenite separated from the tuff. In pure form, both hematite and ilmenite are relatively rare in the samples studied and generally show secondary alteration related to the hydrothermal alteration of dacite and andesite tuffs noted above.

Magnetite. Magnetite is quite widespread in the altered rocks of the zeolitized zone but is less widespread in the propylitized rocks. In the zeolitized zone magnetite is present as two varieties, pseudomorphic after hematite and redeposited. Magnetite pseudomorphic after hematite is developed along the replacement front of hematite by pyrite. Redeposited magnetite is observed in very small cavities of strongly zeolitized tuffs. It forms distinctly bounded octahedral crystals with mirrorlike faces showing no corrosion. It is generally associated with laumontite. In the propylitized zone magnetite is present exclusively as pseudomorphs after hematite. The pseudomorphs are most often observed within the latticework zones of pyrite, in the outer parts of the capillary veinlets of these zones.

Anatase. Anatase is found as isolated dipyramidal crystals or irregular growths (Fig. 173). The crystals show combinations of

Table 42. Interlayer Spacings of Brookite

Line no.	d	I	Line no.	d	I
1	3.736	2	14	1.486	3
2	3.445	10	15	1.468	2
3	3.129	2	16	1.429	3
4	2.855	9	17	1.358	2
5	2.455	4	18	1.329	3
6	2.214	3	19	1.234	5
7	1.952	4	20	1.166	1
8	1.877	5	21	1.152	1
9	1.835	2	22	1.118	3
10	1.677	3	23	1.036	3
11	1.649	5	24	1.023	5
12	1.603	4	25	1.006	4
13	1.530	1			

$p\{011\}$ and $z\{013\}$; the pinacoid $c\{001\}$ is less common. The color is black, or steel gray in fragments. The luster is very strong, approaching adamantine. Fragments show distinct cleavage along $\{011\}$. Under the microscope in reflected light the anatase displays no anisotropy, the color is gray, and the reflection is about 16 to 17%. Table 41 gives X-ray data on the anatase.

Brookite. Brookite is present as isolated isometric crystals (Fig. 173) and, less commonly, as tabular crystals in intergrowths with rutile. The crystals show the forms $m\{120\}$, $l\{122\}$, $b\{010\}$, and $e\{111\}$. The first two, equally developed and giving rise to the equidimensional appearance of the crystals, are predominant. The color is black, less commonly dark green to black. The mineral is opaque. The fracture is uneven, sometimes marked by indistinct cleavage.

The brookite is clearly anisotropic in reflected light in an immersion medium, the color light-gray, and the reflection about 20%. Table 42 gives X-ray data on the brookite.

Rutile. Rutile is present in small quantities in altered rocks of the upper part of the propylitized zone. It is usually associated with "leucoxenized" ilmenite. It is segregated as very thin acicular crystals and less commonly forms paniculate growths.

Cuprite. Cuprite is present in the deposit as a product of the technological activity of man. It is precipitated together with chrysocolla as thin lamellae on the surfaces of the silica gel deposits. Sometimes it is found as very small octahedral crystals growing on stalks of grass that got into the silica gel. In this case it is covered with a chrysocolla crust.

Quartz. Quartz in the deposit is of several generations and of several modes of origin. Clear separation of generations and modes of origins of the Pauzhetka quartz is at present a subject for further study.

The largest quartz crystals (up to 1.5 cm) are in the upper part of the propylitized zone, where they fill veinlets 5 to 7 mm thick and irregular cavities 0.5 to 4 cm in diameter. In the quartz crystals filling the cavities, the prism $m\{10\bar{1}0\}$ and the principal rhombohedron $r\{10\bar{1}1\}$ are well developed; the secondary rhombohedron $r\{01\bar{1}1\}$ is a sharply subordinate form. The crystals are transparent to translucent and contain considerable numbers of gas and liquid inclusions. The quartz is associated with chlorite and calcite. Quartz is just as widespread in the zeolitized zone as in the propylitized zone, but usually it forms very small crystals that do not exceed 5 mm along the principal axis. Here they usually fill irregular cavities and less commonly form thin (about 2 mm) veinlets. The presence of several modes of origin of the quartz within a single mineralized cavity is characteristic. The earliest quartz is represented by long prismatic crystals with equally developed rhombohedra $r\{10\bar{1}1\}$ and $z\{01\bar{1}1\}$. Later generations are presented by short prismatic forms with unequal development of the rhombohedra (the principal rhombohedron predominates) and finely acicular spherulitic aggregates. The earlier quartz is associated with adularia, stilbite, chalcedony, calcite, and less commonly prehnite. The later quartz is associated with heulandite, laumontite, mordenite, calcite, and opal.

Chalcedony. Chalcedony is abundant but less widespread than quartz. It is most common in the upper part of the propylitized zone, as veinlets in rounded almond-shaped cavities. Chalcedony in the veinlets is associated with quartz and calcite; chalcedony in the cavities is associated with quartz, calcite, anhydrite, chlorite, and pyrite. The chalcedony of the veinlets is often recrystallized to granular quartz, with the characteristic spherulitic and fibrous textures of the chalcedony preserved as relicts in isolated quartz grains.

Chalcedony is less common in the zeolitized zone. It fills irregular cavities and is associated with quartz and laumontite.

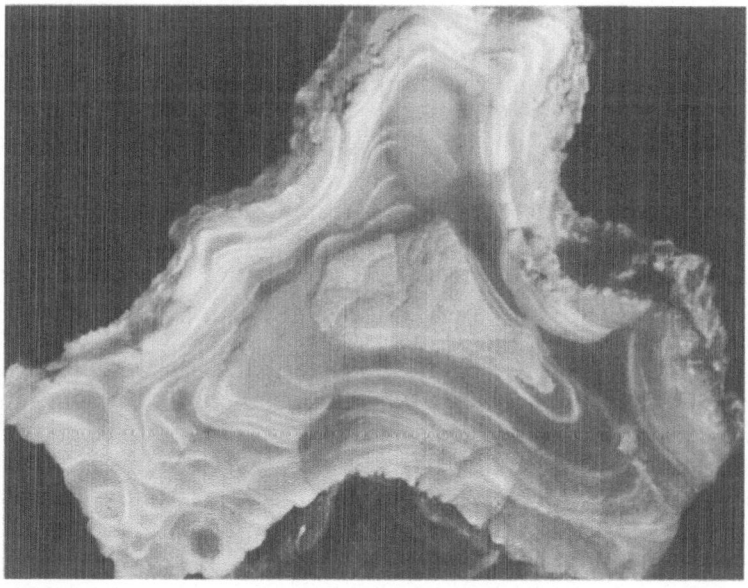

Fig. 174. Conchoidal-reniform forms of chalcedony from a cavity in alluvium (× 5).

Table 43. Chemical Composition of Sinter from the Old
Structure

Component	%	Molecular proportions	Component	%	Molecular proportions
SiO_2	69.62	1.159	MgO	0.43	0.010
TiO_2	0.51	0.006	CaO	2.50	0.045
Al_2O_3	7.66	0.075	Na_2O	1.70	0.027
Fe_2O_3	0.91	0.012	K_2O	1.00	0.022
FeO	0.17	0.003	H_2O^-	6.06	–
MnO	0.02	–	H_2O^+	9.56	–
		Sum		100.14	–

Chalcedony is common in alluvium, where it cements boulders of volcanic rocks. Reniform and conchoidal varieties, covering walls of irregular cavities in the alluvium, are present in addition to continuous dense chalcedony (Fig. 174). Here it is closely associated with vesicular opal, mordenite, stilbite, and laumontite. Chalcedony also commonly forms discoidal and rounded segregations in dense calcic montmorillonite and acts as cement in alluvium.

Opal. Opal is extremely widespread in the deposit as various types of sinter, represented by both old geyser structures and recent sinter in the form of small sinter areas overgrowing crusts and flow-deposited forms of stalk type. Naboko (1954) observed old sinter structures on the left bank of the Pauzhetka River a short distance from Lenivii Geyser. The writer has studied the old geyser structures of First Paryashchii Spring. Only one-third of the structure, on the northeast side of the spring, has been preserved. Most of it has been destroyed by drilling work at this site. In exterior form the structure is a gentle cone with a large (up to 1 m in diameter) outlet in the center. Its greatest thickness is about 25 cm. The larger, lower part (17 to 18 cm) of the structure consists of alternating layers of dense white sinter and friable accumulations of organic remains (chunks of wood, stalks of grass, leaves, etc.) that have been replaced by silica. The exceptionally fine scale of the replacement is remarkable; the finest structural details of the surfaces of the grass stalks are preserved. The upper part (about 7 cm thick) consists of very dense yellowish-gray sinter. Chips of this sinter are translucent. The surface of the structure is irregularly bumpy, porcelanous, and in places enamellike. Under the microscope the dense yellowish-gray sinter is isotropic. At high magnifications a large number of very small (approximately 0.005 mm), distinctly anisotropic acicular crystals can be observed in the overall isotropic mass.

Table 43 gives the chemical composition of this sinter. The analysis was made in the chemical laboratory of the Institute of Volcanology (Siberian Section of the Academy of Sciences of the USSR) by I. M. Bender. From the rather high content of aluminum, calcium, and potassium and the very small anisotropic inclusions observable under the microscope, it can be hypothesized that unidentifiable zeolite minerals are present as an admixture in the sinter.

At the site of the destroyed structure, close to the outlet, new sinter is being formed. Uneven and stalklike aggregates of friable powdery sinter grow on the surfaces of various objects (rock fragments, broken bottles, paper). The formation of these aggregates is related to the strong liberation of a steam phase from the hot-spring solutions, causing an enormous number of very small drops of the solution to be in chaotic motion in the air around the spring. These drops settle on the surfaces of objects from various angles and evaporate there to form specific stalklike sinter deposits.

Accumulations of recent sinter as sinter fields were observed by the writer around the mouth of First Teplyi Creek and in the northern part of the hot-spring area at the divide between the Pravaya Pauzhetka and Levaya Pauzhetka rivers. The sinter fields are surfaces

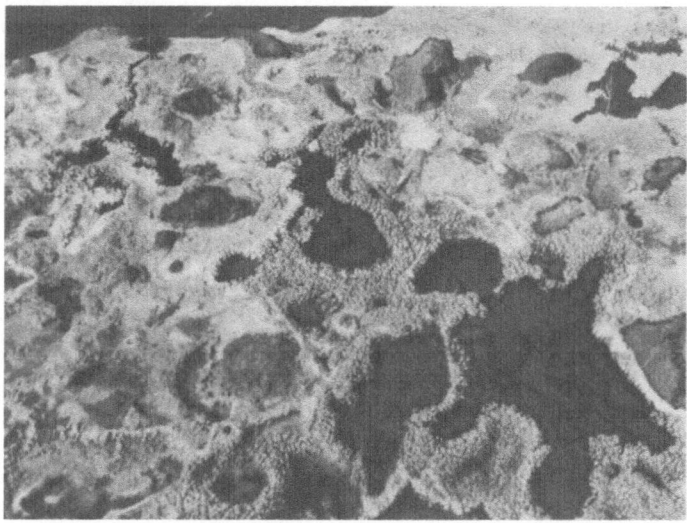

Fig. 175. Sinter from a modern sinter field at the divide between the the Pravaya Pauzhetka and Levaya Pauzhetka rivers (x 1/10).

Table 44. Chemical Composition of Present Sinter

Component	%	Molecular proportions	Component	%	Molecular proportions
SiO$_2$	85.35	1.421	CaO	1.20	0.021
TiO$_2$	Trace	—	Na$_2$O	0.55	0.008
Al$_2$O$_3$	1.45	0.015	K$_2$O	0.19	0.002
Fe$_2$O$_3$	0.08	—	H$_2$O$^-$	5.18	—
FeO	1.96	0.26	H$_2$O$^+$	3.74	—
MnO	0.02	—	S	0.17	—
MgO	0.21	0.005			
			Sum	100.10	—

slightly elevated above the ground level composed of blocks of tuff tightly cemented by sinter. On these surfaces there are many boiling springs and interconnecting pools (Fig. 175). These pools are generally fringed by openwork corallike sinter. Table 44 gives a chemical analysis of sinter of this kind studied by Naboko (1954).

From the foregoing we see that in the recent sinters the quantities of aluminum, calcium, and potassium are considerably lower, and apparently their presence can also be explained by inclusions of zeolites (laumontite).

Opal, primarily as dense, white, uniform masses cementing alluvium, is rare in the hydrothermally altered rocks. It is less common as spherical and vesicular forms in voids in alluvium (Fig. 176), closely associated with quartz, chalcedony, and mordenite. In this case it is the youngest mineral and is precipitated on the surfaces of spherulitic mordenite aggregates.

Silica Gel. Silica gel, present in various quantities, is widespread in the deposit in almost all the hot springs and creeks. Also, there are large accumulations in various man-made structures (outlet pipes, radiators, etc.). Because the study of silica gel makes it possible to clarify the specific features of the morphology of particles of the dispersed phase and the patterns of alteration at various stages of aging, it will be described in more detail.

The greatest accumulations of silica gel were observed in First, Second, and Third Teplyi creeks, most abundantly in Second Teplyi Creek. Therefore, the distribution of the silica gel in the creeks will be described using the deposits in Second Teplyi Creek as an example.

Distribution of silica gel in the creeks (on the example of Second Teplyi Creek). Second Teplyi Creek, containing substantial accumulations of silica gel, is a right tributary of the Levaya Pauzhetka River.

Numerous hot springs in the upper and middle reaches feed the creek. Thin openwork sinter crusts are developed above the water level around isolated blocks of tuff. Both in the conditions of formation and in the character of the aggregates, this sinter is similar to sinter described in detail by Naboko from the field close to Lenivii Geyser on the opposite bank of the Levaya Pauzhetka.

Silica gel is regularly distributed in the creek bed. It was not observed in the lower course of the creek. In the middle course, silica gel in the form of gelatinous films covers small gravel fragments and stalks of dry grass blown into the creek by the wind and also forms irregular lumpy accumulations near colonies of blue-green algae. Upward along the creek the size of these accumulations increases considerably; the algal colonies are completely covered and surrounded by dense gelatinous silica gel. The organic acids liberated by the algae apparently promote strong coagulation of the silica gel and thereby accelerate its accumulation. Small islands of silica gel, often adjoining one another, are formed in the creek. In the uppermost reaches of the creek, the entire bed is covered with silica gel.

Because of the considerable denseness of the silica gel, it could be cut for study in cross section. The deposits are 3.5 to 4.0 cm thick in the stream valley 2.5 to 3 m from the stream. There is weak banding in the silica gel by alternation of dark-gray and grayish-white

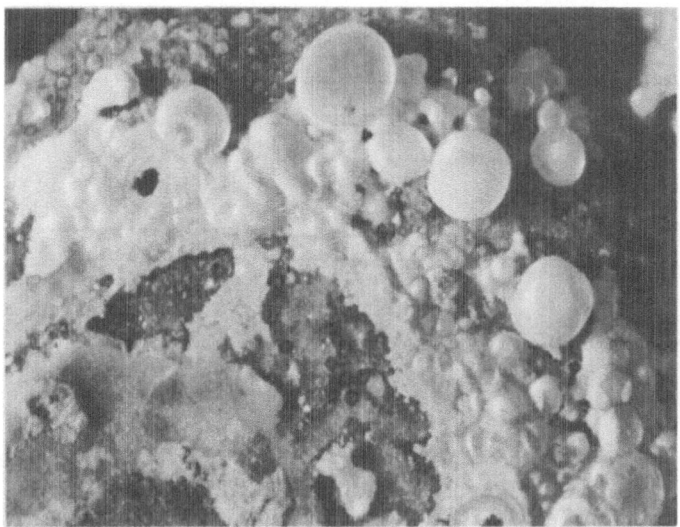

Fig. 176. Spherical and vesicular forms of opal in cavities in cemented alluvium (x 15).

Table 45. Chemical Composition of Dense
Silica Gel

Component	%	Component	%
SiO_2	87.67	Na_2O	0.55
TiO_2	Not determined	K_2O	0.18
Al_2O_3		H_2O^-	0.36
Fe_2O_3	1.34	H_2O^+	6.89
CaO	0.85	Organic	2.39
MgO	0.30		
		Sum	100.53

bands. The dark-gray color, in places almost black, is caused by
decomposed particles of dead algae. The banding is related to
cyclic burial and extinction of blue-green algae. Algae that are
everywhere covered by a uniform dense mass of coagulated silica gel
die. For some time relatively pure silica gel is deposited on the surface of the buried algal colonies. Then new algal colonies are formed
and grow rapidly, and the process is repeated.

Actually, as a rule the deposition surface of silica gel is a dense
crust of white, very pure silica gel. Clear green bunched algal
colonies begin to grow on the crust at individual points. There are

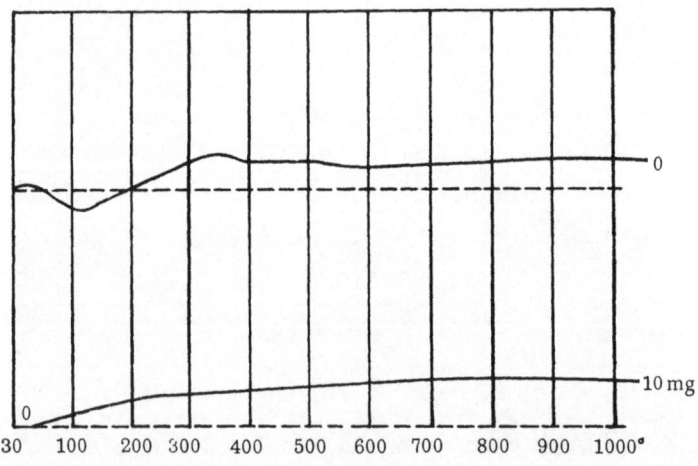

Fig. 177. Heating and weight-change curves of dense silica gel. Sample
153 k/n 925b, March 2, 1962. Weight, 93.4 mg; loss, 9.3 mg (about 10%).

also substantial differences in the consistencies of the differently colored bands of silica gel. The dark varieties colored by organic material are gelatinous. Their viscosity is low, and they can flow. The grayish-white gel is considerably thicker and is slightly brittle but is easily cut with a knife.

Composition and morphology of silica gel aggregates of various consistencies. Under the microscope in immersion preparations the silica gel is observed as irregular lumpy aggregations, and it is isotropic. Globular texture can be seen at highest magnifications. The index of refraction is 1.449. X-ray data show that no crystalline phase is present. This silica gel is amorphous to X-rays.

Table 45 gives a chemical analysis of dense silica gel from the surface crust of the deposits (analysis made by E. P. Ryabichkina). The analysis shows minor aluminum, calcium, sodium, and other elements. The presence of such components as aluminum and calcium in the silica gel is apparently caused by a finely dispersed mechanical admixture of very small laumonite crystals. The molecular proportions of Ca to Al, close to 1:1, supports this hypothesis to some extent. The wide development of laumontite in the rocks of the region that are altered by recent hot-springs waters, where it is the principal component of the thick zeolitized zone, is also evidence in favor of this hypothesis. The presence of sodium is caused by an admixture of sodium chloride, the main component of the thermal solutions,

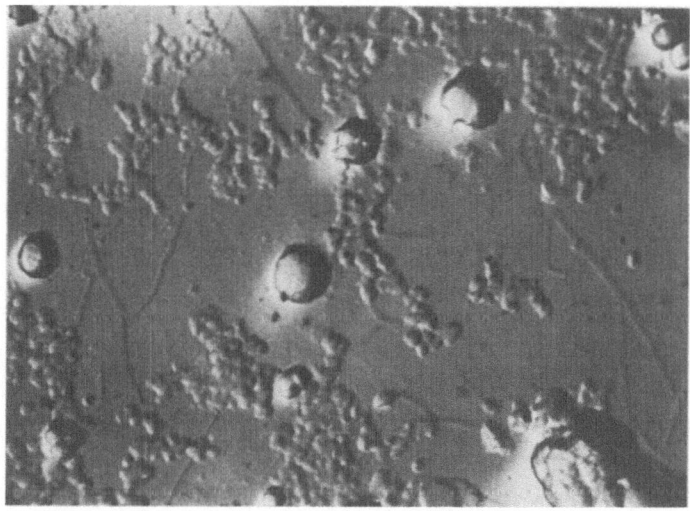

Fig. 178. Nature of particles in silica-gel sol. Carbon replica (x 40,000).

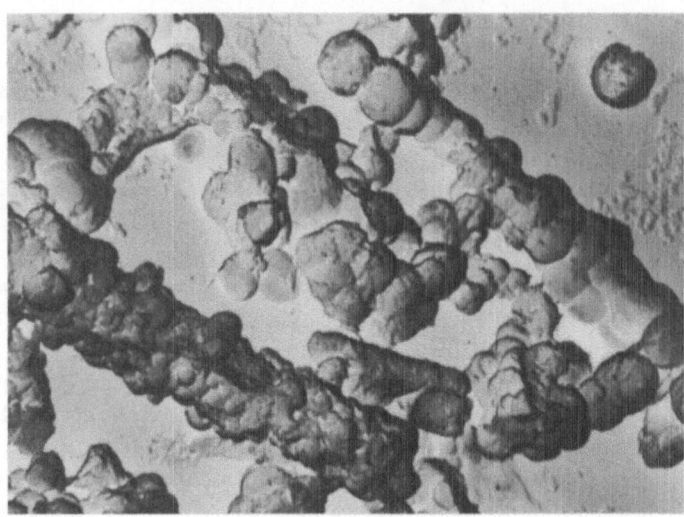

Fig. 179. Chain and pear-shaped aggregates of globules in viscous silica
gels. Carbon replica (× 24,000).

which often forms an efflorescence on surfaces of tuff blocks along
the stream banks. In addition to these main components, the following
elements were identified spectrographically in the silica gel: B (0.n);
Fe (0.n$^-$); Sb (0.0n); Ti, Mn, Zn, As, Ni, Sr, Y (0.0n$^-$); Be (0.00n$^+$);
Ba (0.00n); Cu, Cr (0.00n$^-$); and V (0.000n). There is one endothermic
reaction and one weak exothermic reaction on the heating curve
(Fig. 177), obtained in the Thermal Analysis Laboratory of the Institute
of Geology of Ore Deposits, Petrography, Mineralogy, and Geochemistry. The endothermic reaction is caused by the liberation of
combined water, and the exothermic reaction is apparently caused by
crystallization of the gel. The total weight loss is about 10%, corresponding to the content of water and organic compounds (9.64%) in
the analyzed silica gel.

Both the silica gels, which are of a varying consistency, from
very liquid and gelatinous to thick and solid, and the solution in which
they are found were studied by electron microscopy.

There are a great number of very small silica globules 25 to
50 mμ in size in suspension in the solution. Generally most of the
globules form chain aggregates and irregular accumulations. Globules
in various stages of coalescence are observed in the irregular accumulations. Rare isolated larger globules 0.3 to 0.5 μ in size (Fig. 178)
are formed by coalescence.

The viscous, gelatinous silica gel is an accumulation of chain aggregates of rather large globules 0.1 to 0.4 μ in size (Fig. 179). In some cases these aggregates consist of linear aggregations of individual globules, and in other cases they consist of linear group growths of many globules. In the latter case the aggregates are pear shaped. The aggregates, consisting of linearly growing solitary globules, often branch and form rings and hollow (three-dimensional) structures.

Tubular forms are also commonly observed in the gelatinous gels. These are apparently osmotic filaments of silicon dioxide. As a rule the ends of these membrane tubes have uneven, broken edges. Branching, closed-ended tubes whose ends in one branch are hemispherical are less common. The ends of the offshoots are also rounded. The length of the broken membrane tubes is 1 to 17 μ, and the outer diameter is 0.3 to 0.8 μ. The minimum thickness of the walls is 0.01 μ and the maximum is 0.05 μ.

In isolated cases, longitudinally cleaved tubes showing the inner surfaces, which are relatively even, were observed. Weakly tubercular character is less common. Sometimes small pits about 7 mμ in size could be observed on the inner surfaces of the tubes. These

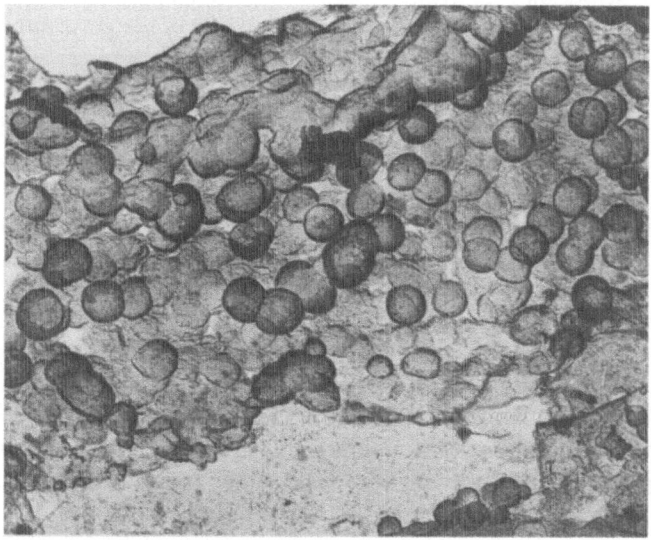

Fig. 180. Uniform globular aggregate of powdery silica gel. In the center of the photograph are irregularly collomorphic forms arising by the coalescence of globules. Carbon replica (× 16,000).

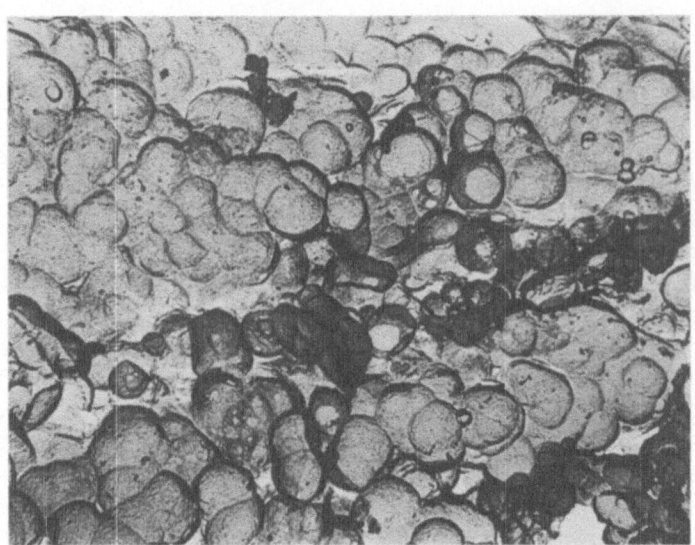

Fig. 181. Microreniform aggregates of powdery silica gel. Carbon replica
(× 18,000).

apparently are relicts of the original pores of the semipermeable membrane filaments of silicon dioxide. From the data of Tinker (1915), who established that the pore size is about 0.16 of the diameter of the particles forming the membranes, we can estimate the probable size of the particles that form these membrane tubes. The computations show that the particle size is in the range of 43 to 45 mμ, which is almost the same as the thickness of the walls of the membrane tubes (50 mμ) in which these pits were noted. Externally the tubes show irregular, often reniform isolated globules and in other cases coalescent globules are often observed on the surfaces of the membrane tubes.

There is an alternation of dense opaline layers with more friable powdery interlayers in the solid silica gel from the surfaces of the crust. The powdery silica gel is a uniform aggregate of relatively large (0.4 to 0.6 μ) globules. There are regions consisting of isolated solitary globules among which pairs of partly coalescent globules and collomorphic forms with irregular outlines formed by the merging of many globules are noted (Fig. 180). Some regions of the powdery silica gel consist of slightly denser microreniform aggregates, in which there are sometimes polygonal globules that have planar boundaries at points of contact. There are very great numbers of more or less uniformly distributed very small (0.025 μ) tubercules on the surfaces of the hemispheroidal elements (Fig. 181). At high magnifications of the electron microscope it was established that these tubercules are vesicular forms in the near-surface part of the gel forming the reniform elements (Fig. 182). These forms apparently reflect the position of particles of water, which is present in these forms owing to the specific conditions of obtaining the replicas (vacuum of 10^{-4} mm Hg).

Crystalline particles are sometimes observed among the globular mass of powdery silica gel. These are intergrowths of wedge-shaped crystals (Fig. 183) very similar in appearance to the laumontite crystals forming crusts on fragments of altered tuffs in the upper parts of the zeolitized zone (well no. 13). The size of the crystalline particles ranges within wide limits from 7–9 to 40–50 μ in diameter.

The opaline varieties of solid silica gel are very dense aggregates of coalescent globules. In general outlines, the structural features of the powdery varieties are repeated in these opaline varieties, and all the morphological types of globular aggregates described for the powdery silica gels are found in them. The difference is the more complete coalescence of globules and the greater denseness of the aggregates. Coalescence of globules and increase in density leads to weakening of the globular texture. There are signs of conchoidal fracture on fracture surfaces of the opaline varieties of silica gel.

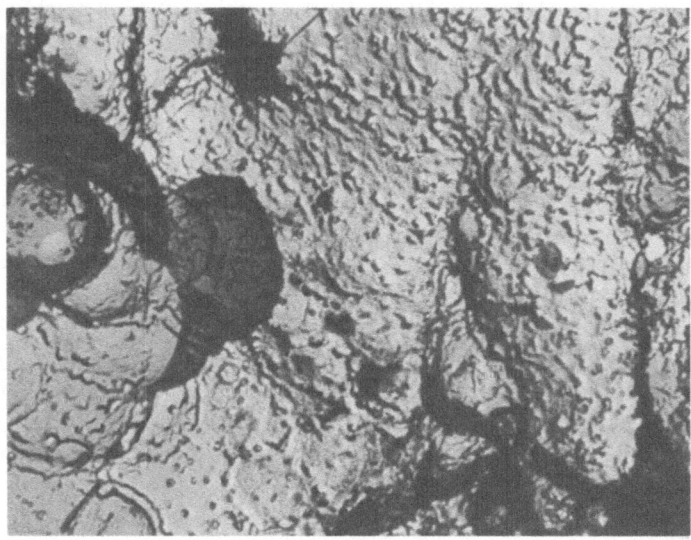

Fig. 182. Distribution of water in the surface region of reniform elements of powdery silica gel. Carbon replica (× 36,000).

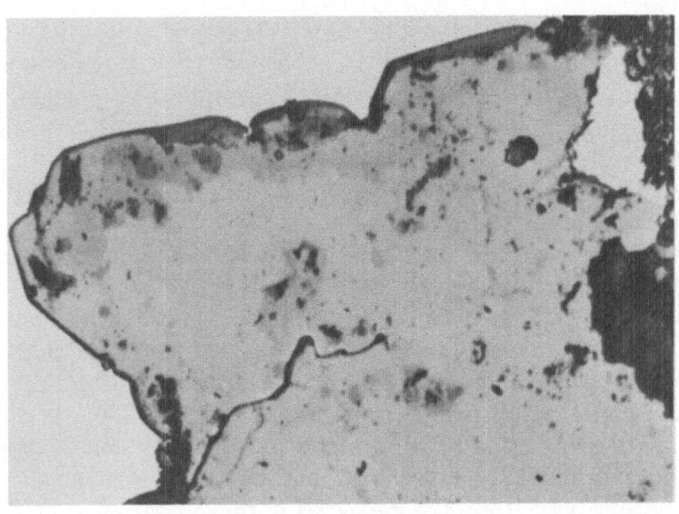

Fig. 183. Crystalline phase in the globular mass of the silica gel. Carbon replica (x 3000).

Thus, study of silica gel of various consistencies under the electron microscope reveals the morphological features of the silica-gel particles and their aggregates that are characteristic of each variety.

The very small (25 to 50 mμ) silica-gel particles found in solutions containing silica-gel coagulates apparently correspond to silica micelles. The existence of these micelles, as aggregates composed of tens and hundreds of particles, is evidence of incipient coagulation. The solutions themselves can be considered highly dilute sols.

Globular texture is present in all the silica-gel coagulates regardless of the consistency. The incipient stages of diagenesis of the silica-gel coagulates are expressed as growth and coalescence of globules, formation of aggregates (chain, pear-shaped, and reniform), and thickening of these aggregates in the transition to opaline varieties. Formation of polygonal globules is also associated with the thickening.

The interrelationship in morphology and genesis between globular and more-complex microcollomorphic aggregates should be especially emphasized. During diagenesis microreniform and other collomorphic aggregates of the silica gel form globular aggregates without crystallization playing any part.

Below we shall briefly describe several silica gels, different in composition, that accumulate in radiators in rooms of the Pauzhetka Observation Station and chrysocolla aggregates that form in the silica-gel deposits as a result of the technological activity of man.

Recent deposition of chrysocolla in the silica gel deposits of Second Teplyi Creek. Though as a mineral species chrysocolla does not belong to the oxide group, nonetheless, considering its features of genesis and place of formation (Second Teplyi Creek), it should be described here.

The writer observed chrysocolla in the middle reaches of Second Teplyi Creek, where it has formed bluish-green film precipitates on silica-gel surfaces. The source of the copper was coils of copper wire that had been put into the creek. Chrysocolla as films 1 mm thick is developed around the coils within a radius of about 11 cm. As thin crusts it also often covers dry stalks of grass blown into the silica gel by the wind.

Under the microscope in immersion liquids the chrysocolla is blue with a greenish cast. It is isotropic and its index of refraction is 1.538. Study of the chrysocolla under an electron microscope showed that its aggregates consist mainly of amorphous particles (globules and globulites) which by clumping together form micro-reniform and pear-shaped aggregates (Fig. 184). The size of the globulites is 25 to 75 mμ, and the size of the globulites is 0.1 to 0.2 μ.

Signs of crystallization of the globular chrysocolla aggregates are observed. In the main globular mass there are regions in which

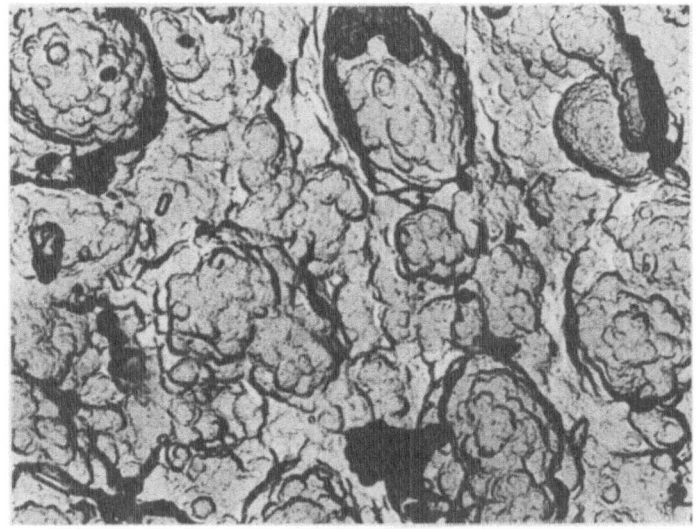

Fig. 184. Globular-reniform structure of a dense crust of chrysocolla. Carbon replica (x 36,000).

Fig. 185. Region of chrysocolla with spherulitic texture. Carbon replica
(x 36,000).

some parts of isolated reniform elements consist of partially coalescent globules and other parts consist of radially oriented crystalline
particles.   Isolated spherulites with distinct radial structures are
observed in these same regions (Fig. 185). The size ranges from 2
to 3 μ in diameter; the crystals forming the spherulites are 1 to
1.5 μ long and 0.1 to 0.2 μ across. Geometrically, these particles are

Table 46. Chemical Composition of Chrysocolla

Component	%	Molecular proportions	Atomic proportions
SiO_2	50.33	0.838	$0.838 = Z$
Al_2O_3	1.42	0.014	$0.028 = A$
Fe_2O_3	1.24	0.007	$0.014 = B$
CuO	24.91	0.392	$0.392 = D$ (Cu)
MgO	1.27	0.031	$0.031 = D$ (Mg)
CaO	2.58	0.046	$0.046 = D$ (Ca)
Na_2O	1.08	0.018	
K_2O	0.56	0.006	$0.048 = E$
H_2O^+	12.06	0.669	
H_2O	3.90	0.216	
Sum	99.35		

distorted hexagons.   There is always void space between the radially oriented crystals in the outer parts of the spherulites.   It should be borne in mind, however, that crystallization of the chrysocolla is far from complete, because this chrysocolla is amorphous to X-rays.

Table 46 gives the chemical composition of this chrysocolla (analysis made in the chemical laboratory of the Institute of Volcanology by I. M. Bender). The analysis was recalculated by the method of Ross and Hendricks (1945), assuming structural analogy between chrysocolla and montmorillonite.   The recalculation coefficient was computed to be 4.854, and the sum of ions in octahedral coordination, 2.71.   The content of exchangeable bases was taken tentatively to be 0.33 ($X_{0.33}$).   Computation of the quantity of Al in tetrahedral coordination leads to $y = 0$ and reveals a slight excess of silica. This is in agreement with visual determination of a mechanical admixture of 5 to 6% powdery silica forming a very thin layer in the central part of the chrysocolla crust.   The computed formula of the chrysocolla corresponds to the following composition:

$$Al_{0.14}Fe_{0.07}Cu_{1.90}Mg_{0.15}Ca_{0.22}(Na,K)_{0.23}Si_{4.06}O_{10}(OH)_2 X_{0.33}$$

$$\Sigma = 2.71$$

Thus, this chrysocolla can be placed in the category in which almost all the aluminum in the octahedral positions is replaced by copper.

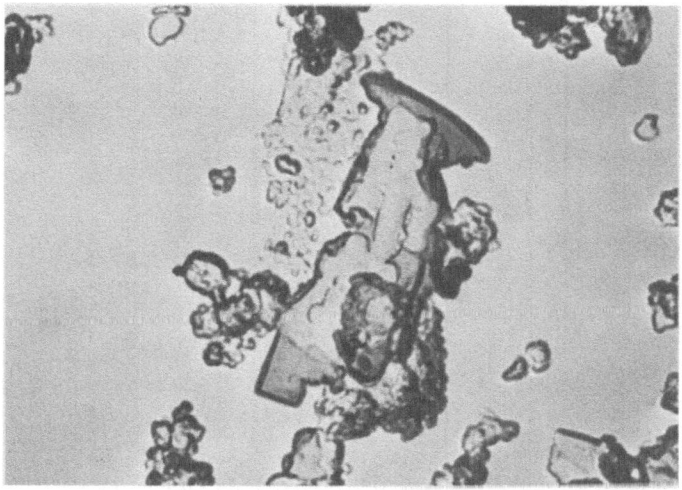

Fig. 186. Skeletal laumontite crystal and small adularia crystal forming in the silica gel from radiators.   Replica with suspension (x 24,000).

Table 47. Chemical Composition of Gel from Radiators

Components	%	Molecular proportions	Components	%	Molecular proportions
SiO_2	63.86	1.064	K_2O	3.84	0.080
TiO_2	0.31	—	H_2O^+	3.92	—
Al_2O_3	14.60	0.143	H_2O^-	2.51	—
Fe_2O_3	3.11	—	CO_2	Not observed	—
FeO	1.14	—	Cl	0.16	—
MnO	Not observed	—	F	Not observed	—
MgO	1.41	—	Sum	99.52	
Ca	3.25	0.059			
Na_2O	1.41	—	$-O = C$	0.04	
				99.48	

Table 48. Interlayer Spacings of Crystalline
Phases from Gels Accumulating in Radiators

Line no.	d	I	Line no.	d	I
1	9.251	6	23	1.694	1
2	6.863	5	24	1.665	1
3	5.068	1	25	1.614	3
4	4.651	2	26	1.538	3
5	4.203	9	27	1.514	2
6	3.688	2	28	1.456	2
7	3.445	5	29	1.431	2
8	3.326	10	30	1.402	1
9	3.252	9	31	1.370	5
10	3.00	5	32	1.340	1
11	2.855	4	33	1.301	1
12	2.748	3	34	1.257	5
13	2.581	6	35	1.227	3
14	2.473	3	36	1.195	3
15	2.332	1	37	1.180	5
16	2.246	2	38	1.143	1
17	2.138	7	39	1.076	1
18	1.976	5	40	1.067	1
19	1.928	2	41	1.043	3
20	1.851	2	42	1.036	2
21	1.799	1	43	1.017	1
22	1.780	7	44	0.989	1

Lines 1, 2, 3, 6, 7, 9, 11, 17, 18, 20, 25, 28, 33, 34, 35, 37, 38, 39,
41, 42, 43, and 44 correspond to laumontite; lines 4, 5, 8, 9, 10,
12, 13, 15, 16, 19, 22, 23, 24, 26, 27, 29, 30, and 32 correspond to
adularia. A number of lines were not identified.

Brief data on silica gel accumulating in radiators. Under an electron microscope, a considerable admixture of cryptocrystalline phase was found in silica gel accumulating in radiators in the living quarters of the Pauzhetka Observation Station. The silica gel has a uniformly globular texture. The size of the globules is 0.03 to 0.1 μ. Isolated globules are rare; generally the globules form pear-shaped aggregates of various shapes and lumpy aggregations.

Two sharply different morphological types of particles form the crystalline phase in the gels. Particles of the first type are elongated, somewhat flattened, and for the most part skeletally developed (Fig. 186). Their average size is 1.5 μ in length and 0.3 to 0.5 μ across. Crystal particles of the second type have a short prismatic and equidimensional habit. The outlines of some of them are reminiscent of feldspar crystals. Their size is 0.2 μ or less. Irregular forms, hackly, spicular, and wedge-shaped with uneven fracture, varying from 0.1 to 5-6 μ in size, are also observed.

Table 47 gives a chemical analysis of this silica gel (made in the chemical laboratory of the Institute of Volcanology by I. M. Bender). The considerable content of aluminum, calcium, potassium, and iron is noteworthy. The molecular proportions of the first three components agree well with the formulas of laumontite and adularia. The presence of iron scale accounts for the iron content.

For final identification, the crystalline phase was separated from the gel by repeated washing. The magnetic fraction was separated from the crystalline phase using an electromagnetic separator, and then the remaining, nonmagnetic fraction was X-rayed with the results given in Table 48.

Thus, minerals characteristic of the authigenic zeolite zone in hydrothermal alteration of dacite tuffs are formed in silica gels accumulating in radiators.

Silicates

Silicate minerals are widespread in the hydrothermally altered rocks of the deposit. However, though the number of observed minerals is considerable, only four are widespread: Laumontite and celadonite in the zeolitized zone and delessite and iron-rich erinite in the propylitized zone. All the others are considerably less common, and some were found only once.

Adularia. According to Naboko, adularia is widespread in the deposit. The greatest quantity of adularia is restricted to the depth interval of 50 to 150 m (upper subzone of the zeolitized zone). According to Naboko, adularia is present in the upper part of this sub-

Fig. 187. Spherulites of sphene in a mass of laumontite (× 150; uncrossed
nicols).

zone as skeletal crystals in cavities leached from plagioclase; in
the lower part the adularia replaces plagioclase phenocrysts. The
adularia is very fresh.   The indices of refraction are: $\epsilon = 1.519$;
$\omega = 1.525$; $\omega - \epsilon = 0.006$; $2V = 68\text{-}70°$ (Naboko's data).   Present-day
deposition of adularia in radiators in the rooms of the Pauzhetka
Observation Station was noted above.

Albite. The writer observed albite in cement in alluvium, where
it fills thin branching veinlets in whose swellings well-formed albite
crystals are present.   The size of the crystals varies from tenths of
a millimeter to 1 mm. The crystals have a tabular habit. The forms
{010}, {001}, and {110} are observed; the form governing the habit
is {010}.   Simple albite twins are widespread. The mineral is
associated with calcite and mordenite.

Prehnite. Prehnite was observed as very small (0.1 to 0.2 mm)
grayish-green crystals in a swelling of a calcite—laumontite vein in
the ower part of the zeolitized zone. The indices of refraction are:
$\alpha = 1.615 - 1.618$; $\beta = 1.620 - 1.623$; $\gamma = 1.643 - 1.647$.

Sphene. Sphene is generally rare in the altered rocks but quite
abundant in the lower parts of the zeolitized zone and the upper parts
of the propylitized zone.   It is present as two varieties: (1) cryptocrystalline aggregates in leucoxene fringes of ilmenite and (2) redeposited sphene, as spherulites and, less often, individual crystals.
The sphene is formed by decomposition of ilmenite and titanaugite.
It often replaces ilmenite pseudomorphically, together with brookite

and anatase. Spherulitic forms of sphene are precipitated in cavities along with chlorite. Spherulites of sphene are often observed in laumontite veins (Fig. 187), where they are restricted to the regions near the boundaries; less commonly they form independent veinlets filled with closely packed spherulites. Also, very small spherulites of sphene are often unevenly disseminated in the rock. Owing to insufficiently pure material, only 3 mg of sphene could be separated for X-ray work; Table 49 gives the results. All the principal lines in the X-ray photograph correspond to sphene. The small number of weak lines is from rock-particle impurities.

Gyrolite. Gyrolite is found in the upper part of the propylitized zone, where it is precipitated in small cavities and, less commonly, fills veinlets and forms (along with chlorite and sphene) pseudomorphs after titanaugite. Naboko observed this mineral as a rather thick veinlet (up to 0.5 cm) in the lower part of the zeolitized zone. Gyrolite grows on the walls of fractures as light-green, foliated spherulitic aggregates. A chemical analysis (Table 50; made by L. A. Basharina) shows a composition close to gyrolite. Table 51 gives the interlayer spacings of the gyrolite.

Celadonite. Minerals of the hydromica group are represented by celadonite. This mineral is widespread in the altered rocks of the zeolitized zone and is sometimes found in the upper part of the

Table 49. Interlayer Spacings of Sphene

Line no.	d	I	Line no.	d	I
1	3.55	4	18	1.571	5
2	3.24	8	19	1.537	3
3	2.985	6	20	1.504	6
4	2.88	4	21	1.445	1
5	2.604	10	22	1.426	5
6	2.543	1	23	1.353	4
7	2.377	1	24	1.312	4
8	2.288	7	25	1.281	4
9	2.115	1	26	1.229	4
10	2.07	5	27	1.194	1
11	1.988	1	28	1.152	1
12	1.958	2	29	1.136	4
13	1.877	1	30	1.118	1
14	1.820	3	31	1.111	5
15	1.752	3	32	1.077	5
16	1.712	4	33	1.044	5
17	1.649	7	34	1.015	2
			35	0.9966	2

Table 50.   Chemical Composition of Mineral
from Okenite–Gyrolite Group

Component	%	Component	%
SiO₂	58.17	MgO	0.25
Al₂O₃	2.22	Na₂O	1.01
Fe₂O₃	0.28	K₂O	0.63
MnO	0.25	H₂O⁺	1.83 *
CaO	32.97	H₂O⁻	1.03 *
		Sum	99.64

*Curve of weight loss upon heating gives 5.25%.

propylitized zone.   The mineral is bluish green.   Generally it is
pseudomorphic after dark-mineral components of the tuff.   Also,
cryptocrystalline collomorphic aggregates of celadonite fill capillary
veinlets.   Less commonly it is deposited as radial aggregates in
small cavities. Under the microscope in transmitted light the mineral
is green, with marked pleochroism from bluish green to yellowish
green.   The indices of refraction are $\alpha = 1.607 \pm 0.002$; $\beta = \gamma$;
$\gamma = 1.638 \pm 0.004$; $\gamma - \alpha = 0.031$.   The wide development of celadonite

Table 51. Interlayer Spacings of Gyrolite

Line no	d	I	Line no.	d	I
1	9.25	9	19	1.840	7
2	7.82	2	20	1.815	2
3	6.25	4	21	1.761	2
4	5.07	2	22	1.649	2
5	4.65	4	23	1.607	1
6	4.23	7	24	1.588	1
7	3.76	5	25	1.507	2
8	3.49	6	26	1.474	1
9	3.13	10	27	1.407	2
10	3.00	2	28	1.349	1
11	2.04	1	29	1.318	2
12	2.83	8	30	1.244	1
13	2.63	7	31	1.195	1
14	2.42	3	32	1.166	3
15	2.30	1	33	1.116	3
16	2.21	4	34	1.060	1
17	2.05	2	35	1.044	1
18	1.934	1			

is apparently caused by elevated calcium content in the upper part of the zeolitized zone.

Datolite. Datolite, extremely rare, was observed in a single instance in hydrothermally altered rocks in the upper part of the propylitized zone as a greenish-gray, fine-grained mass completely filling small (0.5 to 0.7 mm), rounded cavities. Its indices of refrac- are $\alpha = 1.628 - 1.630$; $\beta = 1.652 - 1.654$; $\gamma = 1.667 - 1.670$. Table 52 gives the interlayer spacings of this datolite.

Zeolites. The following zeolite minerals are found in the deposit: laumontite, heulandite, wairakite, mordenite, stilbite, and thomsonite. Laumontite is sharply predominant and stilbite is common. The others are much less common.

Laumontite. Two varieties of laumontite are found in the deposit, ordinary crystalline laumontite and metacolloidal laumontite. We shall first describe the crystalline-granular and well-crystallized laumontite.

Table 52. Interlayer Spacings of Datolite

Line no.	d	I	Line no.	d	I
1	6.77	1	26	1.564	2
2	4.77	1	27	1.530	4
3	4.16	1	28	1.506	2
4	3.74	3	29	1.486	1
5	3.42	4	30	1.423	1
6	3.12	10	31	1.363	2
7	2.99	4	32	1.351	1
8	2.86	9	33	1.328	4
9	2.53	7	34	1.309	4
10	2.47	1	35	1.268	4
11	2.41	2	36	1.250	1
12	2.30	1	37	1.232	6
13	2.26	8	38	1.214	2
14	2.20	9	39	1.199	3
15	2.17	1	40	1.188	4
16	2.08	2	41	1.178	1
17	2.00	6	42	1.144	3
18	1.880	10	43	1.123	2
19	1.815	1	44	1.098	3
20	1.778	3	45	1.086	1
21	1.747	1	46	1.071	2
22	1.720	4	47	1.063	2
23	1.673	1	48	1.040	1
24	1.653	10	49	1.036	4
25	1.618	2			

Fig. 188. Prismatic crystals of laumontite with calcite (x 18).

In studying well cores brought up from the zone of high-tempera-ture waters, laumontite is observed on almost all surfaces in alluvial deposits. Here it grows on well-rounded gravel as small acicular crystals. Slightly deeper it is present mainly as cement in dacite tuff. Regions of alteration to laumontite in the rock have irregular, diffuse outlines. Here laumontite contains abundant inclusions consisting of very small relicts of partially replaced rocks. Still deeper the dacite tuff is replaced uniformly throughout the rock by coarse-grained laumontite aggregates. Laumontite is developed as cement in the plagioclase phenocrysts and even in the dark minerals. The quantity of new laumontite varies within wide limits, reaching a maximum of 50% of the rock.

Laumontite also forms independent veins or veins together with calcite and fills cavities of various shapes in the rock. The lau-montite veinlets and laumontite–calcite veinlets vary in thickness and branch strongly. The thickness ranges from fractions of a millimeter to 2 cm and, less commonly, to 5 cm. The laumontite forms dense columnar aggregates in the thicker veins. At swellings there are drusy crusts of prismatic laumontite crystals on cavity walls (Fig. 188). The formation of drusy cavities is characteristic of regions of branching and intersection of calcite–laumontite vein-lets. Flat spherulitic aggregates of laumontite, developed in the plane of the vein, are extremely characteristic in the thinner veinlets (up to 1 mm).

As noted above, laumontite and calcite are the principal components of the veinlets. The calcite is of two generations. First-generation calcite forms isolated rhombohedral crystals and completely fills the central parts of the laumontite–calcite veinlets. Second-generation calcite is represented by rare scalenohedral crystals growing on first-generation calcite. Sphene deposited as spherulites near the vein boundaries is common, though sharply subordinate in the laumonite veinlets. Less commonly the spherulites of sphene are unevenly distributed over all of the laumontite mass.

The laumontite is snow white in both the aggregates and the individual crystals. Because of coarse striations on the faces of the laumontite crystals it did not seem possible to measure the faces goniometrically. The mineral is brittle and has very good cleavage. Its specific gravity, determined pycnometrically, is 2.39 to 2.40.

Table 53. Interlayer Spacings of Laumontite

Line no.	d	I	Line no.	d	I
1	9.74	3	26	1.946	4
2	9.25	7	27	1.785	1
3	6.78	6	28	1.752	2
4	6.18	1	29	1.699	2
5	4.98	2	30	1.620	6
6	4.43	3	31	1.592	1
7	4.10	9	32	1.567	1
8	3.84	2	33	1.517	2
9	3.596	2	34	1.474	4
10	3.47	10	35	1.441	3
11	3.307	1	36	1.367	5
12	3.216	2	37	1.342	1
13	3.16	1	38	1.322	3
14	2.985	4	39	1.301	2
15	2.85	3	40	1.262	2
16	2.77	3	41	1.244	4
17	2.56	3	42	1.227	4
18	2.496	2	43	1.189	2
19	2.425	5	44	1.166	2
20	2.35	3	45	1.153	1
21	2.27	1	46	1.122	1
22	2.206	1	47	1.117	1
23	2.137	6	48	1.085	3
24	2.07	1	49	1.058	1
25	1.946	4	50	1.023	3
			51	1.023	3

Table 54.   Chemical Composition of Pauzhetka
Laumontite

Component	%	Component	%
SiO_2	52.55	MgO	0.85
TiO_2	None	CaO	11.42
Al_2O_3	20.89	Na_2O	0.47
Fe_2O_3	None	K_2O	0.15
FeO	"	H_2O^-	1.89
MnO	"	H_2O^+	12.85
		Sum. . . .	100.74

Under the microscope twinning on {100} is common. The elongation is positive, and the extinction angle $C \wedge \gamma$ ranges from 27 to 30°. The indices of refraction are $\alpha = 1.513 \pm 0.002$; $\beta = 1.523$; $\gamma = 1.524 \pm 0.002$; $\gamma - \alpha = 0.011$. The heating curve corresponds to the standard heating curve of laumontite. Table 53 gives X-ray data on laumontite from the veins.

Table 54 gives a chemical analysis of laumontite made by L. A. Basharina in the chemical laboratory of the Institute of Volcanology. Recomputation of the analysis (Table 55) by the oxygen method leads to an ideal formula of potassic laumontite; minor sodium and potassium (Table 54) were neglected in comparing the formulas. The presence of magnesium, which apparently replaces calcium isomorphically, is worthy of note.

Metacolloidal laumontite forming rose-colored and brick-red chalky and opaline aggregates of various denseness is present in the deposit in addition to well-crystallized laumontite. Segregations of metacolloidal laumontite are observed in zeolitized andesite–dacite tuff breccias restricted to a definite depth interval, which ranges from 181-190 to 257-262 m in the various wells. In studying cores from several wells, the following depth intervals of distribution of metacolloidal laumontite in the hydrothermally altered rocks of the zeolitized zone were found:

Well no.	Depth interval, m	Well no.	Depth interval, m
14	229–252	6	190–194
17	183–196	21	257–262
19	193–199	4	216–223
16	181–199	20	191–201
12	192–194		

The types of metacolloidal laumontite segregations are diverse. Most commonly, metacolloidal laumontite is present as thin branching veinlets formed by the filling of fractures between andesite fragments and dacitic cement. Irregular pocket segregations of this laumontite 1 to 6-7 cm in diameter are no less common. Linearly oriented veinlets 2 to 6 mm thick cutting both the andesite fragments and the cement are less common. Finally there are rounded amygdaloidal segregations of metacolloidal laumontite 0.5 to 2.5 cm in diameter.

As already noted above, laumontite is present as dense opaline varieties. Under the microscope at low magnifications it is isotropic; at the greatest magnifications the aggregates show weak anisotropy. Irregular fractures filled with well-formed laumontite crystals are often observed in the main metacolloidal mass (Fig. 189).

In each type of segregation, near the contacts, inclusions of calcite forming equidimensional grains with signs of skeletal growth are observed in the metacolloidal laumontite. The red color of this laumontite is caused by finely dispersed hematite. The hematite is distributed irregularly in the metacolloidal laumontite and thus colors it unevenly. Diffusional distribution of the coloring matter was

Table 55. Recomputation of the Chemical Analysis of Laumontite

Component	%	Molecular proportions	Atomic proportions of oxygen	Number of oxygen atoms, computed to 12	Atomic proportions of cations	Number of cation atoms
SiO_2	52.55	875	1750	8.10	875	4.05
Al_2O_3	20.89	205	615	2.84	410	1.90
MgO	0.85	22	22	0.10	22	0.10
CaO	11.42	203	203	0.93	203	0.93
Na_2O	0.47	7	7		14	0.06
K_2O	0.15	1	1	0.03	2	0.01
H_2O^+	12.85	714	820	—	—	—
H_2O^-	1.85	106				
Sum · · ·	100.74		$\begin{array}{r} 3418 \\ - \ 820 \\ \hline 2598 \end{array}$	12		

Overall divisor 2598/12 = 216

Amount of water 820/216 = 3.8

Formula: $Ca_{0.9}Mg_{0.1}Al_{1.9}Si_{4.1}O_{12} \cdot 3.8H_2O$.

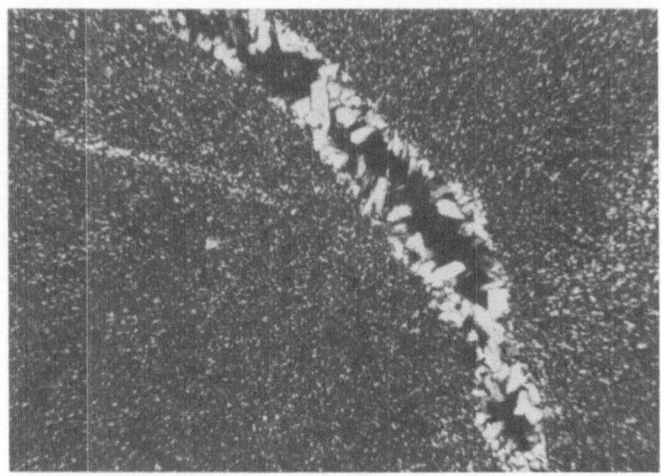

Fig. 189. Weakly polarizing aggregate of metacolloidal laumontite.
The fissure is filled with crystalline laumontite (x 200; crossed nicols).

observed (Fig. 190). Table 56 gives the interlayer spacings of the metacolloidal laumontite.

In conclusion, it should be noted that metacolloidal laumontite segregations in fractures and pores of the rock are localized within the zone of steam formation, which according to Aver'ev (1961) has very complex outlines and, depending on the degree of fracturing of the rocks, ranges from 120 to 200 m and deeper.

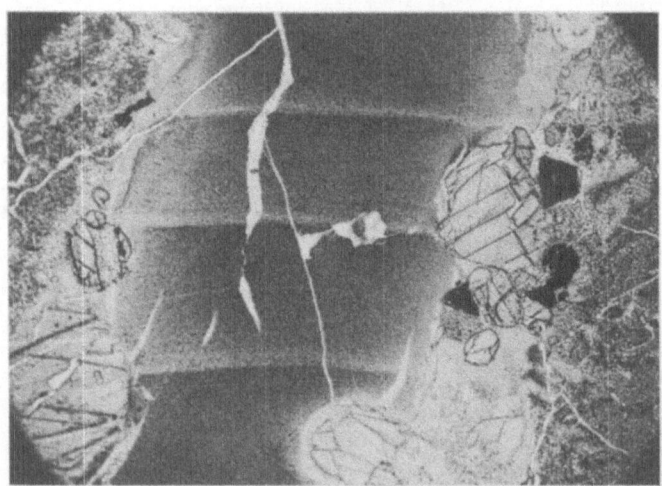

Fig. 190. Diffusional distribution of hematite pigment in metacolloidal
laumontite (x 46; uncrossed nicols).

Heulandite.   Heulandite predominates in the hydrothermally altered rocks of the zeolitized zone. In isolated cases it is observed in the upper part of the propylitized zone. It forms very thin fringes in the contact regions of chalcedony and calcite veinlets. These fringes consist of aggregates of fine heulandite crystals of flesh-red color. The crystals are poorly formed, but cleavage along the pinacoid on {010} is distinct. In the upper part of the propylitized zone, heulandite was observed in irregular cavities as larger crystals (up to 1.5 mm across) associated with calcite and an unidentifiable chlorite mineral.

The indices of refraction of the heulandite are: $\alpha = 1.486 - 1.490$; $\beta = 1.496 - 1.498$; $\gamma = 1.498 - 1.504$. Table 57 gives X-ray data for

Table 56.   Interlayer Spacings of Metacolloidal Laumontite

Line no.	d	I	Line no.	d	I
1	9.10	10	31	1.686	1
2	6.70	8	32	1.665	1
3	6.18	2	33	1.607	5
4	4.97	2	34	1.588	1
5	4.36	3	35	1.553	2
6	4.02	10	36	1.536	1
7	3.86	2	37	1.507	4
8	3.61	2	38	1.503	1
9	3.44	9	39	1.477	1
10	3.30	2	40	1.426	7
11	3.21	3	41	1.356	1
12	3.12	2	42	1.318	2
13	2.97	5	43	1.295	4
14	2.82	3	44	1.271	1
15	2.76	3	45	1.257	8
16	2.69	1	46	1.222	8
17	2.52	4	47	1.182	4
18	2.47	2	48	1.156	5
19	2.42	7	49	1.141	4
20	2.33	4	50	1.115	2
21	2.24	1	51	1.095	2
22	2.16	2	52	1.084	8
23	2.13	8	53	1.075	1
24	2.06	2	54	1.058	1
25	1.958	2	55	1.043	8
26	1.934	4	56	1.036	1
27	1.845	2	57	1.023	5
28	1.825	2	58	1.013	1
29	1.775	3	59	1.006	1
30	1.752	5	60	0.993	3

Table 57. Interlayer Spacings of Heulandite

Line no.	d	I	Line no.	d	I
1	3	11.08	39	3	1.569
2	4	8.94	40	2	1.530
3	2	8.04	41	4	1.498
4	3	5.98	42	2.	1.487
5	1	5.88	43	3	1.456
6	5	5.20	44	5	1.442
7	2	4.89	45	5	1.419
8	5	4.71	46	2	1.401
9	5	4.40	47	3	1.390
10	10	3.94	48	3	1.375
11	1	3.74	49	5	1.360
12	4	3.44	50	2	1.345
13	2	3.31	51	2	1.324
14	2	3.20	52	4	1.307
15	3	3.17	53	2	1.289
16	1	3.10	54	2	1.282
17	10	2.93	55	4	1.270
18	7	2.81	56	3	1.256
19	7	2.74	57	2	1.232
20	2	2.71	58	2	1.196
21	4	2.54	59	3	1.169
22	6	2.44	60	3	1.150
23	1	2.38	61	3	1.133
24	2	2.09	62	2	1.123
25	3	2.03	63	2	1.114
26	2	1.982	64	2	1.104
27	5	1.964	65	3	1.087
28	3	1.869	66	2	1.073
29	2	1.830	67	2	1.065
30	2	1.802	68	1	1.049
31	4	1.775	69	3	1.037
32	4.	1.738	70	2	1.028
33	3	1.703	71	2	1.010
34	1	1.675			
35	2	1.657			
36	2	1.643			
37	2	1.622			
38	3	1.599			

the heulandite. In view of the restricted material, a finely lamellar mineral (celadonite?) intergrown with the heulandite was not separated out.

Wairakite. The writer observed wairakite in a number of cases in the uppermost parts of the propylitized zone. It is rigorously restricted to the depth interval of 250 to 280 m. The largest crystals were observed in one of the core samples of well no. 17, from 256 m, consisting of chloritized andesite tuff.   A rather large cavity of irregular shape consisting of three interconnecting chambers filled with wairakite, laumontite, anhydrite, and calcite was observed in this sample.   The wairakite forms drusy crusts on the walls of the cavity (Fig. 191).   Crystals 0.2 to 0.6 cm in size are translucent and transparent and show combinations of tetragonal trisoctahedra {211} and cubes {100}.   The tetragonal trisoctahedron is the predominant form.

Under the microscope the wairakite is isotropic and does not even show twinning; this distinguishes it from New Zealand wairakite for which Steiner (1955) noted distinct anisotropy and complex twinning. The index of refraction of this wairakite is 1.498 ± 0.002; the specific gravity, determined pycnometrically, is 2.28.

A chemical analysis of the Pauzhetka wairakite (Table 58) made in the chemical laboratory of the Institute of Volcanology by I. M. Bender showed that compositionally it is close to the New Zealand wairakite.   Table 59 gives the chemical analysis recomputed to 100% after subtracting the Fe_2O_3, which was caused by a mechanical ad- mixture.   Table 60 gives the interlayer spacings of the Pauzhetka and New Zealand wairakites.

Fig. 191. Wairakite crystals precipitated on the walls of a cavity in chloritized andesite tuffs (× 10).

Table 58. Chemical Composition of Wairakite

Component	%	Molecular proportion	Component	%	Molecular proportion
SiO_2	53.53	0.891	CaO	11.56	0.207
Al_2O_3	25.09	0.246	Na_2O	1.03	0.032
Fe_2O_3	0.79	0.010	K_2O	Not observed	—
MgO	0.54	0.012	$H_2O\pm$	7.52	0.394
Sum · · · ·	100.06		Sum · · ·	100.06	

Wairakite, as thin crusts in the parts of the wairakite–calcite veins near the contacts, is very rare in the zeolitized zone. The crystals show combinations of {211} and {100}, but the cube faces sharply predominate.

Mordenite. Mordenite is rare. It is most abundant as cement in alluvium, where it was observed as finely acicular radial and tangled fibrous aggregates closely associated with quartz, opal, and stilbite. In small quantities it is present, together with laumontite and calcite, in very small cavities of hydrothermally altered rocks from the zeolitized zone. In the upper layers of the propylitized zone it was

Table 59. Recomputation of Chemical Analysis of Wairakite

Component	%	Molecular proportion	Atomic proportion	Atomic proportion of cations	Coefficients
SiO_2	54.18	902	1804	902	31
Al_2O_3	25.40	249	747	498	17.1
MgO	0.55	015	15	—	0.51
CaO	11.70	209	209	209	7.18
Na_2O	1.04	016	16	32	1.13
H_2O	7.13	384	384	768	13.20
Sum	100.00		2791		

Overall divisor: $2791/96 = 29.07$

Formula: $Na_{1.13}Ca_{7.69}Al_{17.1}Si_{31}O_{96} \cdot 13.2H_2O$ → Pauzhetka wairakite

$Na_{1.12}Ca_{61}Al_{13.35}Si_{27.65}O_{96} \cdot 14H_2O$ — New Zealand wairakite*

*Recomputed analysis of wairakite from Wairakei (New Zealand) taken from Steiner (1955).

observed as inclusions of fan-shaped, finely acicular aggregates in vein quartz.

Stilbite. Stilbite was observed in the zeolitized zone in cavities filled with quartz and calcite. It forms rose-yellow lamellar crystals

Table 60. Interlayer Spacings of Wairakite

Line no.	Pauzhetka		New Zealand		Line no.	Pauzhetka		New Zealand	
	d	I	d	I		d	I	d	I
1	6.78	4	6.85	40	38	—	—	1.722	40
2	6.05	4	—	—	39	1.720	9	1.732	5
3	5.57	10	5.57	80	40	—	—	1.708	5
4	4.73	5	4.84	40	41	—	—	1.696	2
5	4.43	1	—	—	42	1.677	3	1.680	5
6	4.00	1	—	—	43	—	—	1.660	10
7	3.74	5	—	—	44	1.607	2	1.612	5
8	3.62	5	3.34	30	45	—	—	1.595	20
9	—	—	3.42	60	46	1.574	4	1.586	10
10	3.39	10	3.39	100	47	1.480	3	1.487	10
11	3.20	4	3.21	5	48	1.451	3	1.465	5
12	3.02	3	3.04—3.06	10	49	—	—	1.437	—
13	2.88	10	2.909	50	50	1.423	3	—	20
14	—	—	2.897	30	51	1.402	8	1.407	—
15	—	—	2.783	10	52	1.367	1	—	10
16	2.74	4	2.770	10	53	1.349	5	1.354	10
17	2.65	5	2.68	40	54	1.337	5	1.343	—
18	—	—	2.67	10	55	1.301	4	—	—
19	—	—	2.50	5	56	1.279	4	—	—
20	2.475	5	2.489	40	57	1.268	4	—	—
21	2.40	4	2.418	30	58	1.257	1	—	—
22	2.33	2	2.35	5	59	1.227	1	—	—
23	2.26	1	2.26	10	60	1.211	8	1.215	20
24	2.20	5	2.215	50	61	1.173	1	—	—
25	—	—	2.17	5	62	1.166	2	—	—
26	2.13	—	2.147	10	63	1.159	1	—	—
27	—	—	2.115	10	64	1.153	1	—	—
28	2.07	2	2.095	5	65	1.140	1	—	—
29	2.03	3	1.996	20	66	1.128	1	—	—
30	1.988	1	—	—	67	1.111	4	—	—
31	—	—	1.93	5	68	1.103	2	—	—
32	—	—	1.886		69	1.084	1	—	—
33	1.877	6	1.895	30	70	1.075	2	—	—
34	—	—	1.867	10	71	1.067	2	—	—
35	—	—	1.857	30	72	1.058	2	—	—
36	1.835	5	1.844	10	73	1.050	2	—	—
37	—	—	1.822	5	74	1.032	1	—	—
					75	1.018	1		
					76	1.014	4	—	—

Fig. 192. Stilbite crystals from a cavity in zeolitized dacite tuffs. Spherical segregations of opal are seen on the faces of some of the crystals (x 25).

not exceeding 2.5 to 3 mm in size. The forms {010}, {001}, and {110} predominate; the lamellar habit is determined by the sharp development of the second pinacoid {010}. Stilbite was mainly precipitated after quartz and before calcite. It is sometimes noted in cavities in cemented alluvium closely associated with calcite, opal, and mordenite (Fig. 192). It is optically negative; $2V = 0$, which is characteristic of stellerite. The indices of refraction are: $\alpha = 1.484 - 1.487$; $\beta = 1.488 - 1.491$; $\gamma = 1.495 - 1.497$. Because of the fine intergrowth of stilbite and quartz, stilbite could not be separated in pure form. Therefore, X-ray work was done on a mixture of quartz and stilbite (Table 61).

Thomsonite. Thomsonite is present in the deeper layers of the zeolitized zone as dense, radial aggregates completely filling small (0.5 to 1.5 mm) cavities. Individual well-formed crystals of thomsonite up to 1 mm in size were observed in some larger cavities filled mostly by quartz. The indices of refraction of the thomsonite are: $\alpha = 1.527 - 1.530$; $\beta = 1.532 - 1.534$; $\gamma = 1.548 - 1.550$.

Chlorite Minerals

Minerals of the chlorite group are widespread in the hydrothermally altered rocks of the propylitized zone. They are of several species, of which delessite, penninite, and a rare variety, calcic chlorite—erinite, were identified with certainty.

Delessite is precipitated as fine, dark-green (sometimes black) films along the contacts of calcite, quartz, and chalcedony veinlets,

Table 61. Interlayer Spacings of Stilbite and Quartz

Line no.	d	I	Line no.	d	I
1	11.23	2	41	1.809	2
2	10.02	4	42	1.785	4
3	9.09	8	43	1.731	2
4	6.99	1	44	1.699	4
5	5.41	2	45	1.667	5
6	5.14	2	46	1.643	4
7	4.85	3	47	1.594	7
8	4.67	8	48	1.565	2
9	4.48	4	49	1.553	6
10	4.27	4	50	1.515	2
11	4.05	10	51	1.498	2
12	3.72	4	52	1.469	2
13	3.49	2	53	1.445	5
14	3.39	6	54	1.426	2
15	3.34	2	55	1.413	2
16	3.18	5	56	1.390	2
17	3.10	2	57	1.377	3
18	3.03	9	58	1.360	5
19	3.00	1	59	1.336	3
20	2.88	1	60	1.318	2
21	2.83	1	61	1.303	6
22	2.78	7	62	1.277	4
23	2.70	2	63	1.257	3
24	2.61	3	64	1.239	6
25	2.56	4	65	1.225	2
26	2.49	2	66	1.200	4
27	2.45	2	67	1.188	2
28	2.35	3	68	1.178	1
29	2.32	1	69	1.164	5
30	2.27	1	70	1.139	2
31	2.23	2	71	1.125	3
32	2.20	1	72	1.117	2
33	2.13	2	73	1.108	2
34	2.10	2	74	1.092	3
35	2.06	3	75	1.067	4
36	2.04	4	76	1.060	2
37	1.958	1	77	1.045	1
38	1.897	4	78	1.035	2
39	1.864	2	79	1.015	3
40	1.825	3	80	1.001	3

and it also forms fringes in cavities filled by these minerals. It is also observed as series of latticework zones, in which it forms fine independent veinlets and pseudomorphically replaces pyroxene. Sometimes delessite as radial aggregates completely fills small cavities in the altered rocks. The index of refraction is 1.602 ± 0.003. Under the microscope in transmitted light the delessite is green, less commonly rose colored.

Penninite was observed in the deeper parts of the propylitized zone as thin independent veinlets. The indices of refraction are: $\epsilon = 1.566 \pm 0.002$; $\omega = 1.570 \pm 0.002$; $\omega - \epsilon = 0.004$. Erinite, filling numerous rounded cavities, is present in the upper part of the propylitized zone. It is clear blue with very strong pleochroism. The index of refraction is 1.610.

Carbonates and Sulfates

Despite the considerable distribution and great morphological diversity of minerals of these groups, we shall not devote space to their description. No metacolloidal varieties were observed among minerals of these groups, and the nature of aggregates of crystallization origin and morphological features of individual crystals are of interest only for surveys of the mineralogy of the deposit as a whole.

Conclusion

The wide distribution of metacolloids in ores and other mineral aggregates bespeaks the importance of colloids in mineral and ore deposition.

In the main, the specific conditions under which colloidal solutions are formed during ore deposition are not yet clear, and it is not entirely clear that sharp changes of the parameters of the system (and this is perhaps one of the determining factors of the hydrothermal process in general) will inevitably lead to the formation of a colloidally dispersed phase at various stages of the process. Moreover, colloidal solutions can apparently form at the very beginning of the process, at the time of separation from the ore-bearing source, in connection with the separation mechanism itself.

In 1952 Betekhtin formulated and laid the groundwork for a number of important conceptions bearing upon the causes of the movement of hydrothermal solutions. One of the most important conceptions is that of a sharp pressure drop caused by fracturing. Betekhtin noted that a vacuum should be produced immediately after separation of the walls of a closed fracture. Of course, this vacuum would not be maintained for long, but it causes a corresponding reaction of the medium so as to equalize the enormous pressure drop. Above all, this brief vacuum mobilizes various sorts of pore solutions and film solutions, which naturally tend to move toward the fracture. It is apparently by this means that the main part of the water of the ore-bearing solutions (predominantly vadose in nature) is mobilized.

On the other hand, when fractures penetrate into regions of accumulation of "hydrothermal" differentiates, these latter infiltrate into the fractures by ejection. As with any ejection, there is a sharp dispersion of the material into colloidally dispersed phases, and an aerosol is formed. The concept of movement of mineral material by aerosols during endogenic ore deposition was first proposed by Chukhrov (1950 and 1955). Because the endogenic aerosols in fracture voids are in motion while pore solutions from the wall rocks are invading them, conditions favorable for the interaction of these solutions with the aerosol are created. A considerable part of the dispersed phase of the aerosol coagulates, and the other part, assimilated by the ore solutions invading the fracture void, form complex combined colloidal-and-ionic hydrothermal solutions. Coagulation of the aerosol leads to the formation of a viscous gellike vein filling. This undergoes a number of diagenetic changes and ultimately is crystallized. Metacolloidal aggregates that can retain for a long time the texture and morphology of a gel are formed.

The question of the lifetime of gels of various composition is of considerable interest. Experimental data and observations on natural gels allow us to note the following orders of lifetime:

Composition	Duration
TiO gel	Hours
Carbonate gel	Hours to days
Lead sulfide gel	Days
Aluminum hydroxide gel	Days to months
Zinc sulfide gel	Months to years
Silica gel	Years to millenia

The lifetimes of the gels are determined with other conditions of the crystallization ability of the material being equal. In mixed gels, the gel with the shortest lifetime begins to crystallize first.

Further directions of study in the area of the mineralogy of colloids and metacolloids are diverse. Least clear are the courses of diagenesis of gels of various compositions, beginning with coagulation and ending with crystallization. Such matters as the crystallization of material in gels, the regularities of recrystallization of metacolloids, and a number of other questions have been little studied.

References Cited

Andrushchenko, P. F., 1954, Mineralogy of the manganese ores of the Polunochnoe deposit, Tr. Geol. Inst. Akad. Nauk SSSR, Vol. 150, No. 16.

Arbuzova, S. K., 1954, Collomorphic galenite–sphalerite ores of the Iokun'zh deposit in Tadzhikistan, Dokl. Akad. Nauk Tadzh. SSR, No. 12.

Aver'ev, V. V., 1961, Conditions of heating of the Pauzhetka hot springs in southern Kamchatka, Tr. Lab. Vulkanol. Akad. Nauk SSSR, No. 19.

Baker, G., and Frostick, A. C., 1951, Pisoliths, ooliths, and calcareous growths in limestone caves at Port Campbell, Victoria, Australia, J. Sediment. Petrol. Vol. 21, pp. 85-104.

Barth, T. F. W., 1955, Presentation of rock analyses, J. Geol., Vol. 63, pp. 348-363.

Berestneva, Z. Ya., 1953, On the mechanism of formation of colloidal particles, Abstract of doctoral dissertation.

Berestneva, Z. Ya., and Koretskaya, T. A., 1949, Electron microscope study of SiO_2 sols, Kolloidn. Zh., Vol. 2, No. 11.

Berestneva, Z. Ya., Koretskaya, T. A., and Kargin, V. A., 1950, Electron microscope study of TiO_2 sols and the mechanism of formation of colloidal particles, Kolloidn. Zh., Vol. 5, No. 12.

Berestneva, Z. Ya., Koretskaya, T. A., and Kargin, V. A., 1951, On the mechanism of formation of colloidal particles of aluminum hydroxide, Kolloidn. Zh, Vol. 13, No. 5.

Betekhtin, A. G., 1952, Some considerations on the causes of movement of hydrothermal solutions, Zap. Vses. Mineralog. Obshchestva, Ser. 2, Vol. 81, No. 1.

Betekhtin, A. G., Genkin, A. D., Filimonova, A. A., and Shadlun, T. N., 1958, "Textures and Structures of Ores," Gosgeoltekhizdat.

Boydell, H. C., 1925, The role of colloidal solutions in the formation of mineral deposits, Trans. Inst. Mining Met., Vol. 34, Part 1.

Bradley, W., 1929, Cultures of algal oolites, Am. J. Sci., Ser. 5, Vol. 18, pp. 145-148.

Brodskaya, N. G., 1954, Sediment deposition in lakes of the arid regions of the USSR. Aral Sea., in "Deposition of Sediments in Modern Waters," Izd. AN SSSR.

Cherepanov, V. V., 1951, Some regularities of morphology, structure, and replacement in malachite aggregates from Urals deposits, Zap. Vses. Mineralog. Obshchestva, Vol. 80, No. 3.

Chetverikov, S. D., 1956, "Handbook of Petrological Computations," Gosgeolizdat.

Chukhrov, F. V., 1936, Pyrrhotite and pyrite in the Kerchensk ores and some general questions of the genesis of iron sulfides, Izv. Akad. Nauk SSSR, Ser. Geol., No. 1.

Chukhrov, F. V., 1950, On the possible role of aerosols, hydrosols, and hydrogels in magmatic ore deposition, Izv. Akad. Nauk SSSR. Ser. Geol., No. 6.

Chukhrov, F. V., 1955, "Colloids in the Earth's Crust," Akademiya Nauk SSSR.

Churakov, A. N., 1911, On the structure and growth of tubular stalactites, Tr. Sankt-Peterburgskoe obshchestva estestvoispytatelei, Vol. 35, No. 5.

Ehrenberg, H., 1931, Der Aufbau der Schalenblenden der Aachener Bleizinklagerstätten und der Einfluss ihres Eisengehaltes auf die Mineralbildung, Neues Jahrb. Mineral.,Beil., Vol. 64, A.

Ermakov, N. P., 1940, Geology and polymetallic ores of the Zapadnyi Darvaz, in "Geology and Mineral Resources of Central Asia," Izd. AN SSSR.

Garrels, R. M., 1944, Solubilities of metal sulphides in dilute vein forming solutions: Econ. Geol. Vol.,39, pp. 472-483.

Gotman, Ya. D., 1941, Typical features of cassiterite in tin deposits of the USSR. Tr. Instit. geol. Nauk, Akad. Nauk SSSR, Ser. Mineralog., No. 46.

Grigor'ev, D. P., 1949, Determining the beginning of growth of flow-deposited forms of malachite, Priroda, No. 3.

Grigor'ev, D. P., 1953, On the genesis of flow-deposited or metacolloidal collomorphic aggregates of minerals (in connection with the role of colloids in ore deposition); Zap. Vses. Mineralog. Obshchestva, Vol. 82, No. 1.

Grigor'ev, D. P., 1961, "Ontogeny of Minerals, " L'vov University.

Grigor'ev, I. F., and Dolomanova, E. I., 1951, New data on the crystallography and typical features of cassiterite of various genesis, Tr. Mineralog. Muzeya, Akad. Nauk SSSR, No. 3.

Grigor'ev, I. F., and Dolomanova, E. I., 1956, On tin deposits of transitional type between cassiterite–quartz and cassiterite–sulfide types: Tr. IIM, No. 3.

Gritsaenko, G. S., Aidin'yan, N. Kh., and Butuzov, V. P., 1950, On aidyrlite from the Novo-Aidyrla deposit in the southern Urals: Zap. Vses. Mineralog. Obshchestva, Vol. 79, No. 1.

Hess, F. L., 1929, Oolites or cave pearls in the Carlsbad Caverns, Proc. U. S. Nat. Mus., Vol. 76, Art. 16.

Herzenberg, R., 1936, Colloidal tin ore deposits, Econ. Geol. Vol. 31, pp. 761-766.

Herzenberg, R., 1938, Brunckit (Zinksulphidgel): Zbl. Min., Abt. A, No. 12, pp. 373-375.

Ivanitskii, T. V., 1953, On textures and structures of sphalerite and iron disulfide of colloidal origin. Geol. Nauk Tr. Inst. Akad. Nauk Gruz. SSR, Ser. Mineralog. and Petrogr., No. 3.

Ivanov, V. V., 1961, Principal geological conditions and geochemical processes of formation of hot spring regions of modern volcanism, Tr. Lab. Vulkanol. Akad. Nauk SSSR, No. 19.

Karyakin, L. N., and Kainarskii, I. S., 1954, On the formation of flow-deposited aggregates from a gas phase; Zap. Vses. Mineralog. Obshchestvo, Vol. 83, No. 3.

Kholmov, G. V., 1929, Results of the mineralogical and petrological survey of the Sherlo-vogorsk tungsten deposit (Trans-Baikal) of 1926; Izv. Geol. Kom. Vol. 48, No. 10.

Knopf, A. W., 1916, Wood tin in the Tertiary rhyolites of northern Nevada; Econ. Geol. Vol. 11, pp. 652-661.

Koptev-Dvornikov, V. S., et al., 1960, Paleozoic intrusive complexes of Betpakdala, Tr. Inst. Geol. Rudn. Mestorozhd. Petrogr., Mineralog. i Geokhim. No. 44.

Kormilitsyn, V. S., 1951. An example of mineral aggregates formed in the presence of a colloidal solution in endogenic ore deposition, Zap. Vses. Mineralog. Obshchestvo, Vol. 80, No. 4.

Lazarenko, E. K., 1953, On a cryptocrystalline variety of zinc blende from the vicinity of Truskovets in the Carpathian region, Dokl. Akad. Nauk SSSR, Vol. 90, No. 5.

Lebedev, L. M., 1953, On inclusions in quartz and calcite from Mangyshlak, Tr. Mineralog. Muzeya, Akad. Nauk SSSR, No. 5.

Lebedev, L. M., 1954a, Collomorphic sphalerite and galenite, Tr. Mineralog. Muzeya, Akad. Nauk SSSR, No. 6.

Lebedev, L. M., 1954b, Oolitic metacolloidal sphalerite, Dokl. Akad. Nauk SSSR, Vol. 15, No. 3.

Lebedev, L. M., 1959a, Morphological features of quartz and cassiterite of the Shakh-Shagaila deposit, Geol. Rudn. Mestorozhd., Akad. Nauk SSSR, No. 3.

Lebedev, L. M., 1959b, The mineralogy of the Iokun'zh deposit, in "Problems of Mineralogy, Geochemistry, and Petrography," Izd. AN SSSR.

Lebedev, L. M., 1960, Experimental study of the conditions of formation of globular and oolitic aggregates of zinc sulfide, Zap. Vses. Mineralog. Obshchestvo.

Lebedev, L. M., 1961, Secondary ore minerals in hydrothermally altered rocks around the Pauzhetka hot springs, Tr. Lab. Vulkanol. Akad. Nauk SSSR, No. 19.

Lebedev, L. M., 1963, Natural silica gel from Second Teplyi Creek at Pauzhetka (Kamchatka), Tr. Lab. Vulkanol. Akad. Nauk SSSR.

Lemmlein, G. G., 1951, Healing of fissures in crystals and transformation of the shapes of cavities of secondary liquid inclusions, Dokl. Akad. Nauk SSSR, Vol. 78, No. 4.

Lemmlein, G. G., 1953, On the theory of healing of fissures in crystals and the equilibrium form of a negative crystal, Dokl. Akad. Nauk SSSR, Vol. 89, No. 2.

Levitskii, O. D., 1939, "Deposits of Rare and Minor Metals of the USSR," Izd. AN SSSR.

Levitskii, O. D., 1953, On the problem of the importance of colloidal solutions in ore deposition, in "Principal Problems in the Study of Magmatic Ore Deposits," Izd. AN SSSR.

Liesegang, R. E., 1912, Ein Membrantrümer Achat, Zbl. Mineral.

Lindgren, W., 1933,. "Mineral Deposits." New York, McGraw-Hill Book Company. (Cited in Russian translation.)

Link, G., 1903, Die Bildung der Oolithe und Bogensteine, Neues Jahrb. Mineral., Vol. 16.

Mackin, J. H., 1945, An occurrence of "cave pearls" in a mine in Idaho, J. Geol. Vol. 53, pp. 58-65.

Maksimovich, G. A., 1955, Calcite oolites, pisolites, and concretions of caves and mines, Zap. Vses. Mineralog. Obshchestva, Vol. 84, No. 1.

Malinovskii, F. M., 1955, Sulfide-bearing phosphorites of Podolia, Zap. Vses. Mineralog. Obshchestva, Vol. 84, No. 1.

Mathews, A. A. L., 1930, Origin and growth of the Great Salt Lake oolites, J. Geol. Vol. 38, pp. 633-642.

Mokievskii, V. A., and Semenyuk, S. N., 1962, Skeletal growth of crystals in a viscous medium, Zap. Vses. Mineralog. Obshchestva, No. 2.

Morse, H. W., Donnay, J. D. H., and Ott, E., 1933, Composition and structure of artificial spherulites, Am. J. Sci., Vol. 25, pp. 494-498.

Naboko, S. I., 1954, The Pauzhetka geysers, Byul. Vulkanol. St., Akad. Nauk SSSR, No. 22.

Naboko, S. I., 1961, Modern hydrothermal processes and metamorphism of volcanic rocks, Tr. Lab. Vulkanol. Akad. Nauk SSSR, No. 19.

Naboko, S. I., 1963, "Hydrothermal Metamorphism of Rocks in Volcanic Regions."

Naboko, S. I., and Lebedev, L. M., 1963, Recent hydrothermal formation of laumontite at Pauzhetka, Tr. Int. Vulkanol. Sibirsk. otd. Akad. Nauk SSSR.

Naboko, S. I., and Piip, B. I., 1961, Recent metamorphism of volcanic rocks around Pauzhetka hot springs, Tr. Lab. Vulkanol., Akad. Nauk SSSR, No. 19.

Naboko, S. I., and Sil'nichenko, V. G., 1957, Formation of silica gel in solfataras of Golovnin volcano on Kunashir Island, Geokhimiya, No. 3.

Noll, W., 1936, Synthese von Montmorilloniten; Ein Beitrag zur Kenntnis der Bildungsbedingungen und des Chemismus von Montmorillonit, Chem. Erde, Vol. 10, No. 2, pp. 129-154.

Pankov, V. P., 1951, Collomorphic sphalerite of the Verkhnyaya-Kvaisa lead-zinc deposit in Yugo-Osetia, Zap. Vses. Mineralog. Obshchestva, Vol. 85, No. 1.

Pavlov, N. V., 1956, Hypogene magnetite-hematite oolites from the iron deposits of the Angara-Ilim region, Izv. Akad. Nauk SSSR, Ser. Geol., No. 4.

Pilipenko, P. P., 1934, The problem of the genesis of agates, Byul. Mosk. Obshchestva ispytatelei prirody, Otd. Geol., Vol. 12, No. 2.

Pošepný, F., 1874, "Studie der Erzlagerstätte von Rezbanya in Süd-Ost Ungarn." Budapest.

Pošepný, F., 1902, "The Genesis of Ore Deposits." New York, Am. Inst. Mining, Metallurgical and Petroleum Engineers.

Quinke, J., 1902, Die Oberflächenspannung an der Grenze wässriger Kolloidlösungen von verschiedener Konzentration, Ann. Phys., Vol. 9, No. 13.

Radkevich, E. A., 1951, Genetic types of cassiterite-sulfide deposits. Tr. Inst. Geol. Nauk, Akad. Nauk SSSR, Ser. Rudn. Mestorozhd. Vol. 134, No. 15.

Radkevich, E. A., 1952, On ores of colloidal origin, Izv. Akad. Nauk SSSR, Ser. Geol., No. 2.

Reark, J. B., 1952, An occurrence of artificial oolites, J. Sediment. Petrol., Vol. 22, pp. 239-240.

Rogers, A. F., 1917, A review of the amorphous minerals, J. Geol., Vol. 15, pp. 515-541.

Ross, C. S., and Hendricks, S. B., 1945, Minerals of the montmorillonite group, their origin and relation to soils and clays, U. S. Geol. Surv. Profess. Papers, No. 205-B.

Rozhkova, E. V., and Solov'ev, N. V., 1936, Experimental study of the conditions of formation of pisolitic iron and aluminum ores, Tr. Inst. mineral'nogo syr'ya, No. 3.

Rozhkova, E. V. and Solov'ev, N. V., 1937, On the problem of the formation of oolitic and spherulitic textures, Byul. Mosk. Obshchestva ispytatelei prirody, Otd. Geol, Vol. 15, No. 4.

Schade, H., 1909, Zur Entstehung der Harnsteine und ähnlichen konzentrisch geschichteter Steine organischen und anorganischen Ursprungs, Z. Chem. Indust., No. 4.

Shadlun, T.N., 1942, Collomorphic ore textures in the Yaman-Kasa deposit in the southern Urals, Zap. Vses. Mineralog. Obshchestva, Vol. 71,

Shafranovskii, I.I., and Mokievskii, V.A.,1956, Growth conditions, geometry, and symmetry of skeletal crystals, Zap. Vses. Mineralog. Obshchestva, Vol. 85, No. 2.

Smirnov, S.S., 1947, The present status of the theory of formation of magmatic ore deposits, Zap. Vses. Mineralog. Obshchestva, Vol. 76.

Steiner, A., 1955, Wairakite, the calcium analogue of analcime, a new zeolite mineral, Mineral. Mag., Vol. 30, pp. 691–698.

Sterk, G., 1953, Erzmikroskopische Untersuchungen des Brunckits (Kryptokrystalline Zinkblende), Berg.- und Hüttenmaenn. Monatsh. Montan. Hochachule Leoben, Vol. 98, No. 4.

Strelkin, M.F., 1953, On the problem of greisenization of granites, in "Problems of Petrography and Mineralogy, Vol. 1, Izd. AN SSSR.

Stulov, N.N., 1953, On twinned cassiterite intergrowths, Zap. Vses. Mineralog. Obshchestva, Ser. 2, Vol. 82, No. 1.

Terziev, G.I., 1962, Morphology and genesis of metacolloidal pyrite aggregates in lead–zinc ores from Madan, Izv. Geologicheski institut, Sofia, Vol. II (in Bulgarian).

Tinker, F., 1915, The microscopic structure of semipermeable membranes and the part played by surface forces in osmosis, Proc. Roy. Soc. (London), Ser. A, Vol. 92, No. 641.

Traube, M., 1867, Experimente zur Theorie der Zellenbildung und Endosmose, Arch. Anat., Physiol. und Wiss. Med., Vols. 1 and 2.

Twenhofel, W.H., 1928, Oolites of artificial origin, J. Geol., Vol. 36, pp. 564–568.

Vernadskii, V.I., 1910, "Mineralogy," 3rd. ed., Vol. I. Moscow.

Vernadskii, V.I., 1912, "Mineralogy," 3rd ed., Vol. II. Moscow.

Vernadskii, V.I., 1925, "History of the Minerals of the Earth's Crust," Vol. 1, No. 1.

Vital', D.A., 1948, Modern calcareous–manganese nodules and oolites in lakes of the Kulunda steppe: Byul. Mosk. Obshchestva Ispytatelei Prirody, Otd. Geol., Vol. 23, No. 2.

Volynskii, I.S., 1946, Mineralogy of sulfide concretions of the Novoaidyrla nickel deposit, Zap. Vses. Mineralog. Obshchestva, Vol. 75, No. 3.

Watanabe, M., 1924, Zonal precipitation of ores from a mixed solution, Econ. Geol., Vol. 19, pp. 497–503.

Wherry, E.T., 1914, Variations in the compositions of minerals, J. Wash. Acad. Sci., Vol. 4, No. 5, pp. 111–114.

Zemann, J., 1950, "Brunckit"—kryptokrystalline Zinkblende, Tschermaks Mineral. Petrog. Mitt., F. 3, Vol. 1, No. 4, pp. 417–419.

Zhilinskii, G.B., 1955, "Typical Features of Cassiterites of Central Kazakhstan," Izd. AN Kaz. SSR.

Index

ERRATA

Page 123, lines 14, 18   for *tin* read *tungsten*
Page 214, caption to Fig. 155   for *cassiterite* read *sphalerite*